DROSOPHILA

To E. B. Basden

INVERTEBRATE TYPES

DROSOPHILA

Bryan Shorrocks, B.Sc., Ph.D.

*Lecturer in Zoology,
University of Leeds*

GINN & COMPANY LIMITED, LONDON

U.K.	Pergamon Press Ltd., Headington Hill Hall, Oxford OX3 0BW, England
U.S.A.	Pergamon Press Inc., Maxwell House, Fairview Park, Elmsford, New York 10523, U.S.A.
CANADA	Pergamon of Canada, Suite 104, 150 Consumers Road, Willowdale, Ontario M2J 1P9, Canada
AUSTRALIA	Pergamon Press (Aust.) Pty. Ltd., P.O. Box 544, Potts Point, N.S.W. 2011, Australia
FRANCE	Pergamon Press SARL, 24 rue des Ecoles, 75240 Paris, Cedex 05, France
FEDERAL REPUBLIC OF GERMANY	Pergamon Press GmbH, 6242 Kronberg-Taunus, Hammerweg 6, Federal Republic of Germany

PREFACE

In writing this book it was my aim to provide a brief yet wide-ranging account of the biology of the fruit fly *Drosophila*. I hope it will prove useful to students and staff in sixth forms, colleges, and universities where these flies are cultured and studied.

Although I have attempted to deal with most aspects of *Drosophila* biology, the book is somewhat biased towards ecology and genetics because these are the two fields in which I am most interested. Previous texts have relied largely upon examples from work carried out in the United States. In this book I have tried to redress the balance a little by referring to as much European information as possible. I have included a key to the more common British species which, unlike many other keys to the genus, uses body coloration and pattern extensively. This has been possible because of the many excellent colour plates and line drawings created by Hilary Burn, to whom I am most grateful.

My thanks are due also to Mr E. B. Basden for reading the manuscript and for providing specimens of *D. kuntzei* and *D. deflexa*. For useful comments on the key I should like to thank Mrs Elsie Broadhead and Mr David Burn. Dr H. C. Bennet-Clark kindly placed the photograph of *Drosophila* courtship at my disposal and Professor B. P. Kaufmann the photographs of the salivary-gland chromosomes. Mr Arthur Holliday took the rest of the photographs. Finally, for sharing the burden of typing the manuscript and providing much helpful comment and encouragement throughout the period of preparation I should like to thank my wife.

Bryan Shorrocks

CONTENTS

INTRODUCTION

In the history of biological investigation certain animals have contributed much more to our knowledge of the subject than have others. This book is concerned with just such a group—the flies of the genus *Drosophila*.

An article published in the Proceedings of the American Academy of Arts and Science in 1906 describes one of the first uses of *Drosophila* as an experimental animal:

"The organism used in this experiment on close breeding was *Drosophila ampelophila* Löw, a small dipterous insect known under various popular names such as 'the little fruit fly, pomace fly, vinegar fly, wine fly, and pickled fruit fly'."

The author, W. E. Castle, working at Harvard University in the United States, wanted to investigate the effects of inbreeding and cross-breeding on fertility and variability. Looking around for a suitable experimental animal he was introduced, by a graduate student, to *Drosophila ampelophila (melanogaster)*, a fruit fly already being cultured in the laboratory for embryological work. The experiments on inbreeding were begun in 1901 and Castle's paper published in 1906 was the result.

Other early accounts of *Drosophila* include a paper which appeared in the Transactions of the Entomological Society of London in 1907, in which E. E. Unwin describes the flies in the following manner:

"Vinegar flies, small brownish two winged flies with bright red eyes, belong to the Muscid family. The commonest species is *Drosophila funebris*."

Unwin mentions their attraction to fermenting substances, saying that the larvae live in rotten and fermenting fruit, while the adults frequent beer and vinegar casks. Further evidence of the attractive nature of alcoholic liquids is provided by the observation that:

". . . an open decanter of claret will attract swarms of them in the summer time."

Another early reference is that made by Howard (1900). He included *Drosophila* among common excrement breeders and suggested that they may be carriers of the typhoid bacterium.

Of the early accounts of *Drosophila*, Castle's is perhaps the most important historically. As a result of his work, *D. melanogaster* came to the notice of T. H. Morgan of Columbia University, New York. The fruit fly appeared to be an ideal subject for the genetical investigations being carried out by Morgan and his colleagues. It was easy to maintain in the laboratory (Castle had simply reared his flies on banana placed in glass tumblers covered with a square of glass); the life cycle was short; and a pair of flies gave rise to a large number of progeny.

Drosophila became firmly established during the next few years as an animal highly suitable for laboratory experiments. Many mutant forms made their appearance, arising quite spontaneously among the progeny of normal individuals. These mutations provided a source of variation essential for the further investigation of the mechanism of heredity.

A further important aspect of *Drosophila* study began in 1933 when T. S. Painter made an examination of the salivary-gland chromosomes of *D. melanogaster*. These chromosomes provide excellent material for cytological studies (see Chapter 5) and their use in genetic research has proved to be of considerable value.

Early work with *Drosophila* centred around one particular species— *D. melanogaster*. Later studies made use of a wider range of species. In particular, members of the "obscura" group of *Drosophila* have been examined widely both in Europe and the United States.

After almost 70 years of investigation, the biology of *Drosophila* is now well documented. One aspect remains inadequately explored, however—the investigation of wild populations. Apart from a few notable exceptions, little progress has been made since Unwin's observation in 1907 that an open decanter of claret will attract the flies. The majority of the work on natural populations of *Drosophila* has, in reality, been carried out under laboratory conditions in population cages. Laboratory work of this kind is invaluable, of course, but wild populations must not be ignored. One of the aims of this book is to suggest that fruit flies, apart from inhabiting laboratories, are also available in the wild for further ecological and genetical investigation.

1 GENERAL BIOLOGY

All the information of a general biological nature that cannot conveniently be placed in any later chapter is described here. The account refers mainly to *Drosophila melanogaster*.

External features
One of the most important features present in all insects is the hard, external, protective covering known as the *integument*. It has contributed greatly to the success of the insects as a terrestrial group of invertebrates.

The integument is produced by the epidermis and consists of three layers. On the surface there is a thin layer known as the *epicuticle*. The rest is the *endocuticle* which, in the harder regions of the integument, is partly converted to a deep-brown layer called the *exocuticle*. Although mainly composed of non-living material, the cells of the epidermis do produce fine protoplasmic filaments called *pore canals* which run through the integument almost to the surface.

The most familiar component of the insect integument is the nitrogenous polysaccharide known as *chitin*. It is closely related to cellulose and, like cellulose, has a sub-microscopic structure of tiny crystals or rods. These rods are bound together by a protein matrix and are arranged at random in sheets that can slide over each other. In this state it is mainly found in the endocuticle and is responsible for combining flexibility with toughness. In the exocuticle, the protein which binds the chitin crystals together has been converted into a horn-like substance called *sclerotin* by a process known as sclerotization. This results in the hardness of the integument in such parts as the head capsule and the limbs. Sclerotin also makes it possible for insects to develop wings, another feature which has contributed to their success.

All small terrestrial insects like *Drosophila* have a large body surface in relation to their volume. Thus, they must be protected from excessive evaporation or they will soon become dehydrated. Water-loss is, in fact, reduced by a wax layer in the epicuticle. This battle against water-loss is a major one so far as insects are concerned and many references to adaptations that minimize this loss will be found throughout this chapter. In addition, there are behavioural mechanisms which ensure that insects such as *Drosophila* are active only at those times when conditions are favourable (Chapter 3).

As well as forming a cover for the body surface, the integument is also found on the surface of those internal tubes formed during development by invagination of the body surface. These include the foregut, hindgut, tracheae, and genital ducts.

In many flies, including *Drosophila*, there is one modification of the integument during development that has a protective function. When the last larval stage is fully grown, the epidermis adds a substance to the existing soft integument so that it slowly hardens and darkens to form the shell known as the *puparium*. Within this, the very delicate *pupa* is formed.

The head   The external appearance of the head of an adult *Drosophila melanogaster* is shown in Figure 1. The *ptilinal suture* is the external mark left by the *ptilinum*—a membranous sac pushed out of the head capsule by the pressure of blood when the adult fly is emerging from the puparium.

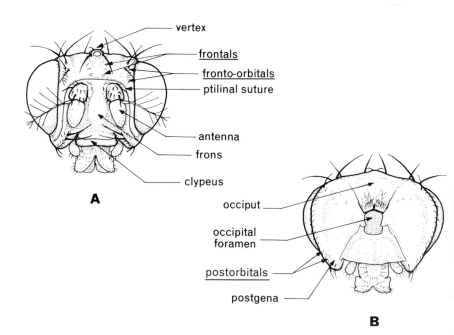

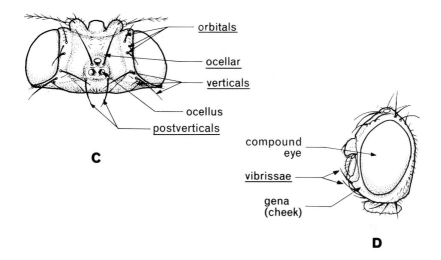

Fig. 1 Head of *Drosophila melanogaster*; **A** anterior view; **B** posterior view; **C** dorsal view; **D** lateral view. Underlined labels refer to hairs and bristles.

It serves to push off the end of the puparium and once this operation is completed it is withdrawn into the head and collapses.

The eyes are of two types, simple and compound. The simple eyes, called *ocelli*, are 3 in number and form a triangle on top of the head. In *Drosophila melanogaster* they are amber. Each consists of a simple lens and an underlying light-sensitive layer of cells—the retina. The ocelli are not capable of even the crudest type of image formation. They are thought to act as stimulatory organs which increase the insect's movement towards light (*positive phototaxis*) by increasing the sensitivity of the brain to light received through the compound eyes. In laboratory experiments the response of fruit flies to light is much more rapid and longer lasting if the ocelli are left intact than if they are blackened.

Light-sensitive organs have two functions: first, an orientating function (*phototactic*) and second, a stimulatory function (*photokinetic*). In fruit flies, the phototactic response is an escape reaction and its efficiency is increased by the photokinetic ocelli.

The compound eyes are more highly specialized light-receivers. Each eye is made up of 680 to 700 cylinder-like units called *ommatidia* (fig. 2) which radiate from the light-sensitive parts of the brain and end at the surface of the eye as a honeycomb of hexagonal lenses.

Light is received by the lens and conducted along the *rhabdome*. The light brings about a chemical change which stimulates the surrounding sense-cells. The points of light received by each ommatidium form a mosaic image of the outside world. The fineness of this pattern of points determines the quality of vision and depends on the number of ommatidia per unit area. In the human eye the receiving elements are the rods and cones of the retina, which are much more numerous per unit area than are ommatia. The difference between the kind of vision we enjoy and that of a fruit fly can be illustrated by examining newsprint with the naked eye and then through a hand-lens. With regard to the perception of different wavelengths of light, *Drosophila* appear to be most sensitive to those of blue-green and ultra-violet.

The function of the pigment in the ommatidial units is to prevent the entry of oblique rays of light which would hinder the formation of a clear mosaic image. Two types of pigment are present in *Drosophila*: a red pigment called *drosopterin*, and a brown pigment called *ommochrome*.

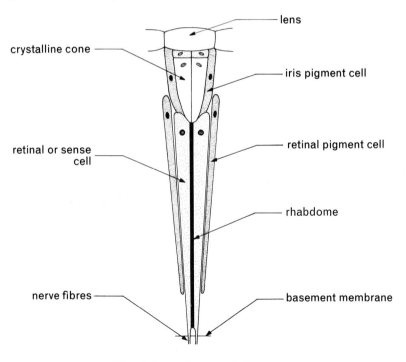

Fig. 2 Insect ommatidium.

The brown pigment is found only in the iris pigment cells, but both occur in the retinal pigment cells (fig. 2). Differences in the amount and distribution of these pigments determine the mutant variations in eye-colour. The surface of the eye is covered with fine hairs that arise in the spaces between each ommatidium.

Compared with many insects, the antennae of *Drosophila* and other related flies are greatly modified (fig. 3). The first segment, the *scape*, is very small and forms a narrow ring around the base. The second segment, the *pedicel*, is larger and somewhat swollen. All the segments beyond the pedicel are called collectively the *flagellum*. The third segment is large and bulbous, and the fourth and fifth are greatly reduced in size and are

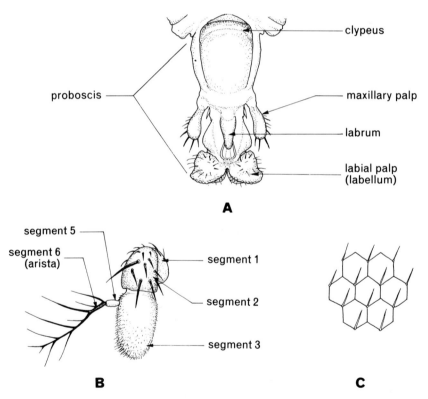

Fig. 3 *Drosophila melanogaster*; **A** mouthparts, anterior view; **B** antenna; **C** surface of compound eye showing facets and bristles.

situated at the base of the sixth segment, called the *arista*. The antennae are well supplied with sense-cells and are thought to play a role in receiving sound (Chapter 6) and chemical stimuli. For example, *Drosophila melanogaster* will orientate towards fermenting fruit at about 40 cm. However, when the antennae are painted over, the flies respond only when very close to the food.

The mouthparts (fig. 3) have been modified for the intake of liquid food. The basic insect mouthparts have become altered to form a fleshy protrudable *proboscis*. At the tip of the proboscis are two large, soft, expansible *labella* (lobes) which have on their surface a series of *pseudotracheae* (collecting channels) that converge at the centre of the lobes. These channels are kept open by narrow sclerotized bands and conduct the liquid food to the tip of the *labrum*. The labrum lies in a groove along the front of the *labium* and forms a tubular canal up which the food is drawn by the sucking action of the pharynx in the base of the proboscis. There are no *mandibles*, and the *maxillae* are reduced to a pair of *maxillary palps*.

The thorax, separated from the head by a flexible neck, consists of three segments. In the basic insect form the three segments are of approximately equal size. They are the *prothorax*, which carries the front pair of legs but no wings, the *mesothorax* and the *metathorax*, both of which bear a pair of legs and, usually, wings.

In *Drosophila* (fig. 4), the prothorax has become greatly reduced and its dorsal portion, the *pronotum*, is merely a collar extending across the

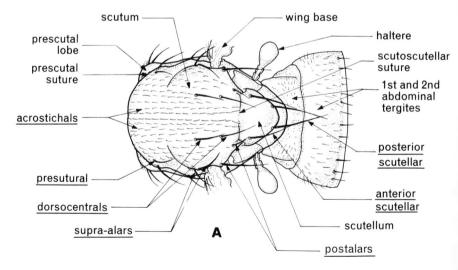

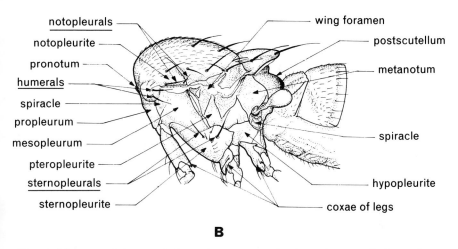

B

Fig. 4 Thorax of *Drosophila melanogaster*; **A** dorsal view; **B** lateral view.

thorax; it is almost concealed beneath the bulging anterior edge of the *mesonotum*—the dorsal surface of the mesothorax. The mesothorax and metathorax have become fused into a solid box called the *pterothorax* and there has been tremendous enlargement of the mesothorax. These changes are probably associated with the fact that the wings on the mesothorax carry the entire flight function, and the metathoracic wings are reduced to *halteres* (balancers; fig. 5).

For any form of flight it is necessary to produce a force that can be used to overcome the effects of gravity and provide, at the same time, movement in the desired direction.

In insects, contracting flight-muscles, using stored food reserves, liberate energy that is transformed through a complex skeletal mechanism into movement of the wings. The wings move the surrounding air and create regions of reduced pressure in front of and above the insect. At the same time, regions of increased pressure appear behind and below it and the insect is driven along this pressure gradient. By taking photographs of the moving wing at extremely small intervals of time, it is possible to plot the wing position throughout one cycle of movement (fig. 6), which lasts about 5·5 msec. Wing-speed is more or less constant through the down-stroke. The up-stroke, however, starts slowly, then speeds up until, finally, the wing rests for about 1 msec at the top of the stroke, vertically above the abdomen. In fact, most insects normally fly with the body at an angle, the head being higher than the abdomen. This is especially so at low

17

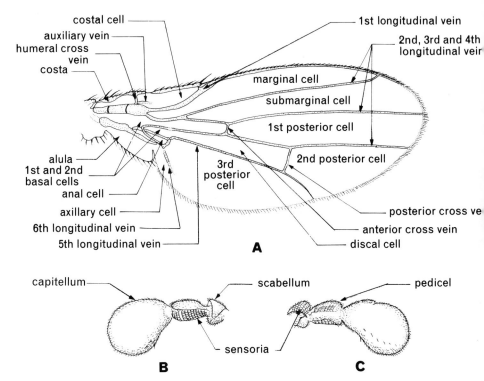

Fig. 5 *Drosophila melanogaster*; **A** wing; **B** haltere, ventral view; **C** haltere, dorsal view.

velocities. When hovering, for example, the plane of vibration of the wings becomes nearly horizontal.

With relatively heavy-bodied, small, winged insects such as *Drosophila*, the wings are vibrated 100 or more times per second under normal conditions. However, there are a number of factors that can change this; age is one. In newly emerged individuals of *Drosophila melanogaster* the wing vibration-rate is relatively low. During the first few days of life the rate rises to a plateau and remains there until the onset of senescence. It is usually higher in males than in females.

Temperature and humidity are also important in determining the wing-beat frequency. In saturated air the frequency increases steadily with an increase in temperature until the death-point is reached at 39 to 40° C. At a relative humidity of 60 to 70% the rate rises until a plateau is reached at 28 to 32° C. It then declines until the thermal death-point is reached.

18

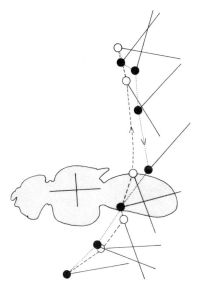

Fig. 6 The wing movement of *Drosophila melanogaster* during flight. The solid circles show the downstroke, open circles the upstroke.

During long periods of flight, fatigue will also influence the rate of wing-beat. Experiments have usually been carried out on fastened flies allowed to fly until exhausted. Under these conditions flights of 1 to 2 hours have been recorded for various species of *Drosophila*. The number of wing-beats measured in these circumstances can be more than a million. The wing-beat frequency usually follows a typical pattern. Initially, the rate fluctuates for some time about a level determined by the species, by the age and sex of the individual, and by the environmental temperature. After some minutes, however, the rate normally declines until it has dropped to a value only half that of the initial frequency. At this point flight usually stops and further stimulation is ineffective. In addition to changes in the wing-beat frequency, the length of the stroke or amplitude can vary also. In fact, such changes are important in turning the animal right or left. The amplitude of either wing can be changed independently, resulting in a pressure gradient which turns the insect from the side of greater amplitude.

Like all aerofoils, the wing of an insect can operate at its best when the flow of air passing over it is streamlined rather than turbulent. This means that in insects which have two pairs of wings, the second pair work at a disadvantage compared with the first because the movement of the first wing introduces turbulence into the departing air-stream. No doubt this is why many of the insects have either lost one pair of wings or developed a mechanism that results in both pairs being closely associated to form a single aerodynamic unit. This latter situation is found in the Hymenoptera (bees and wasps) and the Lepidoptera (butterflies and moths). In the Coleoptera (beetles) the front pair of wings are usually hard and serve to protect the hind pair used in flight. In the Diptera (flies) the hind pair of wings is represented by halteres.

The halteres or balancers vibrate in a vertical direction with the same frequency as the wings, but out of phase with them. Amputation of the halteres does not affect wing-beat frequency, but it does interfere with stability in flight. It has been suggested, therefore, that they act as stabilizers.

The structure of the legs is shown in Figure 7. In many species of *Drosophila* there is a comb, or combs, of enlarged bristles on the first two

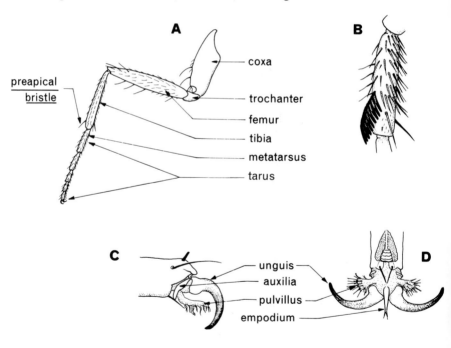

Fig. 7 Leg of *Drosophila melanogaster*; **A** first leg of female; **B** metatarsus of first leg of male showing comb; **C** pretarsus, lateral view; **D** pretarsus, lower view.

tarsal segments of the foreleg of the male. There is some evidence that the tarsi might contain organs of taste. Immediately the tarsi come into contact with a sugar solution the proboscis is extended.

The abdomen   The appearance of the abdomen and the difference between male and female terminalia are shown in Figure 8. Compared with the basic insect form, the segmentation in both sexes has been modified by secondary developments. The seven pairs of *spiracles* (respiratory openings) are assigned to the first seven abdominal segments. In the female, the genital opening is situated between segments eight and nine,

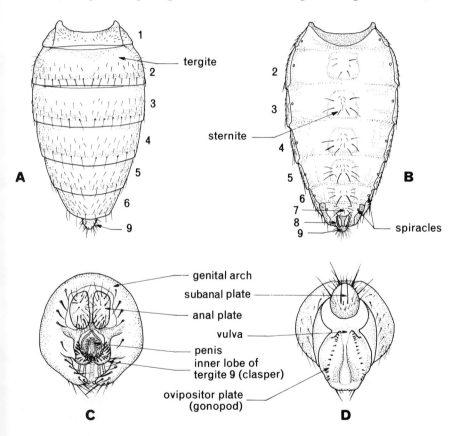

Fig. 8 Abdomen of *Drosophila melanogaster*; **A** dorsal view of female; **B** ventral view of female; **C** terminalia of male; **D** terminalia of female.

concealed by the two ovipositor plates—the *gonopods*. The ninth segment is, in fact, rather small and bears the anus. The tenth and eleventh abdominal segments have become reduced, being represented only by membranes around the anus. In the male, the segmentation has become even more complex. The seventh segment has lost all trace of sclerotization and is entirely membranous. The eighth segment consists only of small plates on either side of the body. The tenth segment is represented by a pair of plates lying alongside the anus, and the eleventh segment is entirely membranous. It is the ninth segment which has become greatly modified. The large and downward-projecting dorsal plate (*tergite*) is known as the *genital arch*. It is continued as a narrow band beneath the anus, from which arises a pair of curved lobes—the *claspers*—which bear a row of stout black bristles called the *clasper comb*. The ventral plate (*sternite*) is an elaborate structure whose posterior edge is extended into several finger-like processes. The middle pair of these encloses the genital opening.

Internal features
The major internal systems of the fruit fly will be described separately for clarity. In reality, each system is intimately associated with the others.
The digestive system Figure 9 illustrates the main parts of this system,

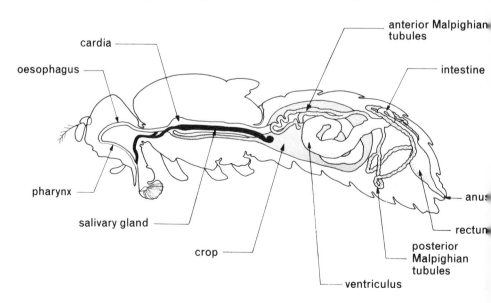

Fig. 9 The digestive system of *Drosophila melanogaster*.

which includes the alimentary canal and its associated organs. Liquid food is drawn into the *cibarium* (pharynx) by a sucking action and, from here, passes into the alimentary canal. This is simply a modified tube that runs from the mouth to the anus. The mouth lies between the *oesophagus* and so-called pharynx which is, in fact, pre-oral. The total length of the alimentary canal is nearly 7 mm, which is 3 to $3\frac{1}{2}$ times the body length of the fly. The canal can be divided into three parts: foregut, midgut, and hindgut. The foregut and hindgut have a chitinous lining but the midgut does not. The foregut includes the oesophagus, *crop* and *stomodaeal valve* within the *cardia* (see fig. 9). The midgut consists only of the *ventriculus*, the anterior end of which forms the cardia. Finally, the hindgut contains the *intestine* and the *rectum*, terminating in the anus. Throughout the system there are specialized valves which help to control the direction of food-movement.

The oesophagus is a narrow tube which extends from the pharynx, through the brain (the *cephalic ganglion*), and into the anterior part of the thorax, where it joins the cardia. The crop is a sac on a slender stem which joins the oesophagus just before the latter reaches the cardia. There are circular muscles at this point which control the flow of food. The saccular part of the crop is bilobed and is capable of great distention. In flies that have only just emerged from the puparium, the crop is unexpanded and its walls are compressed into compact folds. After feeding it might increase to 8 or 10 times its original volume, causing distention of the entire abdomen. When the crop is dissected out in salt solution its colour is found to be pale amber and its walls may show muscular contractions. The food is transferred to the ventriculus for digestion.

The cardia, also called the *proventriculus*, is an oval swelling at the anterior end of the ventriculus and covers the valve formed by the end of the oesophagus. Part of the epithelium of this region secretes a chitinous membrane called the *peritrophic membrane* that is drawn back as a thin cylindrical sheath enclosing the food. The gut of insects contains no mucous glands whose secretions will lubricate and protect the epithelial cells. This function has been taken over by the peritrophic membrane, which is permeable to digestive enzymes and to the products of digestion. The ventriculus, or stomach, is the longest part of the alimentary canal and here the food is digested. It is a thick-walled tube with weakly developed muscles which finally narrows in the *pyloric region* and joins the hindgut. In freshly emerged flies the ventriculus is partly swollen by the swallowing of air. This operation forces blood into the wings and causes them to expand. Later, however, the air disappears into the body by a process of diffusion.

The short and narrow tube which forms the intestine contains the remains of food material still enclosed in the peritrophic membrane. At the point where it joins the rectum there is a valve which is thought to help in drawing the peritrophic membrane backwards. The rectum is a rather thin-walled sac that tapers posteriorly to the anus. Projecting from the side walls into the interior of the rectal sac are two pairs of large conical papillae. It has been suggested that these rectal papillae of insects act as water-absorbing organs. Circular muscles at the anus control the periodic expulsion of waste material from the digestive system.

From the pyloric region of the ventriculus there arises two pairs of long slender *Malpighian tubes*. The posterior pair of tubes is simple and unmodified, but the terminal portion of the anterior pair is thin-walled and bladder-like. The tubes are often distended by white or greenish contents thought to be calcium salts. The Malpighian tubes are excretory organs; their waste products pass out of the body by way of the hindgut.

On either side of the ventriculus are two slender unbranched tubes, the *salivary glands*, extending back through the thorax. The end of each gland tapers into a salivary duct and these unite to form one common duct opening into the labial groove in the proboscis. The exact function of the saliva has not been demonstrated although it may possibly act to dissolve solid foodstuffs before ingestion.

The circulatory system of *Drosophila* (fig. 10) is of the "open" type characteristic of insects. It consists of a dorsal vessel, the *haemocoel*, accessory contractile organs, septa, and the blood or *haemolymph*.

The dorsal vessel is made up of two parts: the posterior heart and anterior aorta. The heart is a four-chambered tube about 1·1 mm long situated close to the dorsal body wall. In the latter half of each chamber there is a pair of *ostia* (slits) with inwardly pointing flaps which act as valves. The posterior end of the heart is closed. The aorta is a narrow thin-walled tube which extends from the anterior end of the heart into the head. The haemocoel is the general body cavity and the blood is the only body fluid surrounding all the internal organs. It is a colourless liquid containing a few free blood-cells.

Circulation occurs in the following way. Blood from the haemocoel is drawn into the heart through the lateral ostial valves and pumped forward by its contractions. Its beat is about 140 per minute. The blood pours into the head from the open end of the aorta to circulate around the brain and proboscis. From here it flows backwards into the thorax and abdomen. The legs are divided throughout their lengths by a membrane and a stream of blood passes down one side of the membrane to the tarsus and then returns up the other side.

Also shown in Figure 10 are two other parts of the internal structure that can be conveniently mentioned here. Serially arranged in a row of

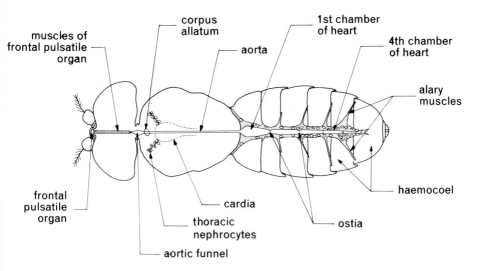

Fig. 10 The circulatory system of *Drosophila melanogaster*.

20 to 25 on each side of the heart and about 10 on each side of the cardia are large oval cells called *nephrocytes*. Their function is not definitely known, although it has been suggested that they remove from the haemolymph materials that are too complex for immediate excretion. Lying on the dorsal wall of the aorta is a single rounded mass of tissue known as the *corpus allatum*. It is an organ of internal secretion (an endocrine gland) regulating growth and reproduction. Certainly its secretions are essential for the development of eggs in females.

The respiratory system The problem of preventing water-loss in insects has already been mentioned in the section on the integument. Another mechanism to combat water-loss is the invagination of the moist respiratory surface into the body cavity where a high humidity can be maintained economically. The spiracles, which are the external openings of this system, can be closed whenever conditions favour excessive evaporation. The branching tubes making up the respiratory system are called *tracheae* and they ramify throughout the body, ending in very small tubes called *tracheoles*. In these final terminal portions of the system, exchange of gases between the air and the tissues can occur. The tracheae are prevented from collapsing by rings of chitin that stiffen the walls. In *Drosophila*, certain parts of the main tracheae are dilated to form thin-walled air-sacs that probably bring about a more efficient aeration of the tissues and may also give buoyancy during flight.

In a small insect such as *Drosophila* the movement of gases through the system of tracheae probably occurs simply by diffusion. The consumption of oxygen by the tissues will produce a lower concentration of this gas in the tracheal endings. An oxygen gradient will be set up, along which fresh supplies will constantly move from the air outside. In a similar way, carbon dioxide will move outwards from the region of high concentration near the tube-endings. Oxygen consumption does not, in fact, stay constant throughout life. It reaches a maximum just before the second larval moult, with a second lower peak just before pupation. During the pupal period consumption drops rapidly, then rises a little to an adult level similar to that seen in the last larval stage. Of course, increased activity brings about an increase in the respiration-rate. During periods of flight it may increase enormously. In *Drosophila repleta*, for example, the respiration-rate during flight is 13 times higher than when the insect is resting. Thus, during periods of sustained flight the spiracles will be fully open to cope with the increased demand for oxygen. At the same time there will be greater water-loss and, therefore, fruit flies are normally active and on the wing only when external conditions are favourable to them (Chapter 3).

The nervous system   In insects the *central nervous system* typically consists of a *supraoesophageal nerve mass* (the brain) connected with a series of segmentally arranged nerve-centres or *ganglia*. These latter structures, together with their paired connections, make up the *ventral nerve cord*. In *Drosophila* (fig. 11), the ganglia have become joined into two masses of nervous tissue located in the head and thorax. No ganglia occur in the abdomen. The nerve mass in the head is made up from the supraoesophageal and suboesophageal ganglia which have become merged together around the oesophagus. The thoracic mass results from the fusion of the three thoracic ganglia and all the abdominal ganglia.

In addition to this central system there is a small *visceral nervous system* consisting of a single elongated ganglion, lying between the aorta and the oesophagus and associated with the gut.

Apart from these nerve centres there are, of course, the peripheral nerves, which are too numerous to mention in detail.

Associated with the nervous system there are the many sense organs that enable the fly to receive and respond to external stimulation. The compound eyes and ocelli have already been mentioned. Since insects are encased in a rigid, mainly non-living integument, special sense organs covering the body and appendages are very important.

Sense hairs and bristles (*setiform sense organs*) are found all over the integument, flexibly attached to the surface and connected to a nerve cell or group of cells at the base. They are known to respond to tactile and

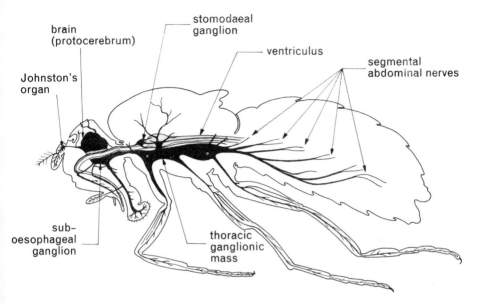

Fig. 11 The nervous system of *Drosophila melanogaster*.

chemical stimulation. Sometimes, the external projection is peg-like and conical. For example, the so-called tasting papillae of the labella are of this type. Experiments indicate that pegs and cones on the antennae detect odour and humidity differences. The dome organs—thin-walled convex domes in the integument attached to single sense cells—are another kind of sensory apparatus. Their function is uncertain although it has been suggested that they detect stress in the integument. Dome organs occur on the wings, halteres, and legs.

The male reproductive system The relative positions of the male organs are shown in Figure 12, where they have been spread apart for clarity. Inside the body the coils of the *testes* are much more compact. All the organs are colourless, or nearly so, in recently emerged flies. Gradually, however, the testes, *seminal vesicles*, and *vasa deferentia* become bright-yellow in *Drosophila melanogaster*.

The testes are filled with *spermatozoa* (male sex cells) in various stages of development; mature sperm are already present when the fly emerges from the puparium. In *Drosophila*, each individual sperm may be as long as 1·75 mm and in one species, *Drosophila hydei*, a sperm-length of 6·64 mm has been recorded.

The anterior ends of the vasa deferentia are swollen to form the

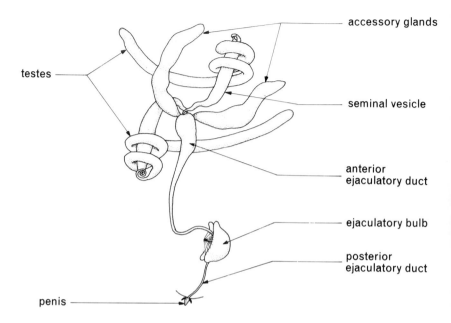

accessory glands

testes

seminal vesicle

anterior
ejaculatory duct

ejaculatory bulb

posterior
ejaculatory duct

penis

Fig. 12 The male reproductive system of *Drosophila melanogaster.*

seminal vesicles. In newly emerged individuals these contain no sperm, but later each is filled with a dense mass of mature sex cells. The anterior ejaculatory duct is about 1·2 mm long in *Drosophila melanogaster.* From the ejaculatory bulb runs the posterior ejaculatory duct, a thin-walled tube that opens to the exterior between the ninth and tenth abdominal segments at the tip of the penis—a chitinized tube differing greatly in shape from species to species. The penis can be pushed out through the genital opening between the lower ends of the genital arch.

In addition, there is a pair of elongated sacs, the *accessory glands,* that open separately into the anterior ejaculatory duct. The glands are filled with a colourless, cloudy fluid which seems to be necessary for the fertilization of the eggs. Sperm that are taken from the seminal vesicles and injected into a female produce fewer offspring than do normally ejaculated sperm injected in a similar way.

The female reproductive system   The various parts of the female system are shown in Figure 13. The *ovaries* are connected by two short lateral *oviducts* to a common oviduct which leads into the pouch-like genital chamber. This chamber opens to the outside through an opening called the *vulva,* which is positioned between the ovipositor plates. This serves a dual function for copulation and as an exit for the eggs. The genital chamber can hold only one egg at a time and, when empty, is contracted with its walls thrown into folds. The ovaries are compact, pyramid-shaped organs, surrounded by fat-tissue. Their size is very variable depending on the extent of development of the eggs and how well fed the female is. When large, they can cause the abdomen to swell greatly. Each ovary is made up of *ovarioles* (egg tubes) varying in number from 10 to 30. Other organs open into the genital chamber: a compactly coiled tube called the *seminal receptacle*, a pair of *spermathecae* (occasionally three), and a pair of accessory glands.

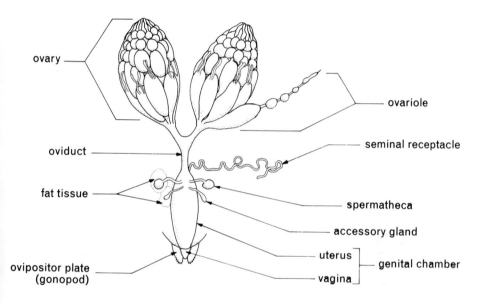

Fig. 13 The female reproductive system of *Drosophila melanogaster*.

When the egg leaves the ovary, it passes down the oviduct by a series of muscular contractions and finally comes to lie in the genital chamber. During copulation, sperm from the male are deposited in the uterus. The female receives approximately 4000 sperm at one mating and these fill first the seminal receptacle and then the spermathecae. This migration of the sperm from the genital chamber to the various storage organs is very rapid. The first sperm appear in the seminal receptacle within a few minutes of ejaculation and this becomes full within an hour. Movement of the male sex cells is thought to be stimulated by a secretion from the accessory glands. When the egg reaches the genital chamber, sperm enter from one of the storage organs and one brings about fertilization. Sperm stored in the seminal receptacle are used first; those from the spermathecae later. Some degree of embryonic development may occur while the egg is still inside the female, but it is finally deposited by contraction of the muscles of the uterus and posterior abdominal segments.

2 LABORATORY ECOLOGY

Ecology is the study of the interrelations of living organisms with their environment. *Drosophila* have been studied mainly in one particular environment, the laboratory, and much has been written on this aspect of the fly's ecology.

Life-cycle of *Drosophila melanogaster*
In the laboratory, fruit flies are usually kept in 250 cm³ (half-pint) milk bottles containing food (see p. 109), and stoppered with cotton-wool.

The adult female lays her eggs on the food surface. These hatch into larvae which moult twice, giving three larval *instars* (stages). The last instar forms a puparial case by hardening its outer skin and, within this, it changes into an *imago* (adult). The length of time between the egg stage and the emergence of the adult varies according to the temperature.

The effect of temperature is important in laboratory studies because, by keeping the flies at different temperatures, one can control the day on which the adults emerge.

temp.°C	Days taken to complete stages of development		
	egg/larval period	pupal period	total time
20°C	6·3	6·3	12·6
25°C	5·0	4·2	9·2

The eggs of *D. melanogaster* (fig. 14) are about 0·5 mm in length. At one end is a pair of filaments which have two functions. First, they prevent the egg from sinking too far below the surface of soft food; second, they are respiratory organs. The surface of these filaments consists of an open network of air-filled spaces. The outer membrane of the egg, called the *chorion*, consists of two layers. The outer layer is impermeable to oxygen;

31

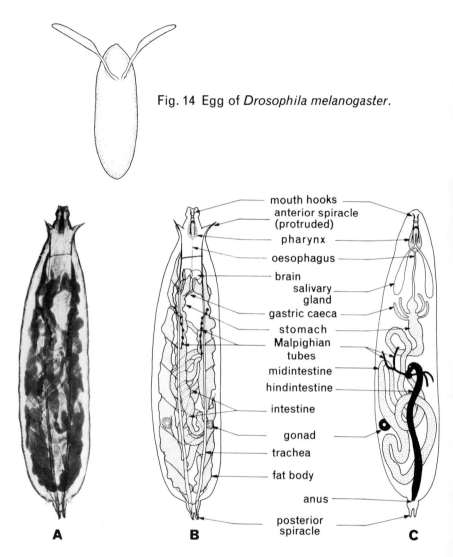

Fig. 14 Egg of *Drosophila melanogaster*.

mouth hooks
anterior spiracle
(protruded)
pharynx
oesophagus
brain
salivary
gland
gastric caeca
stomach
Malpighian
tubes
midintestine
hindintestine
intestine
gonad
trachea
fat body
anus
posterior
spiracle

A   **B**   **C**

Fig. 15 Third instar larva of *Drosophila melanogaster*; **A** photograph
of dorsal aspect; **B** interpretative drawing of photograph;
C simplified diagram showing arrangement of larval organs.
Note the sizes of the gonads (black circle, male; white spot,
female).

between the layers lies a film of air continuous with the air-spaces in the filaments. These project above the surface of the food and provide a direct route for the entry of oxygen into the layer of air held within the chorion. In more natural environments the filaments often become submerged in a surface layer of water. Under these conditions the dissolved oxygen from the surrounding water is obtained by diffusion, with the filaments acting as a kind of gill.

At 25° C the first moult occurs after about a day, the second after about 2 days, and the formation of the puparium about 4 days after hatching. In the final stage—the third instar—the larva may reach a length of about 4·5 mm. The larval body is made up of 12 segments: 1 head segment, 3 thoracic segments, and 8 abdominal segments. The mouth is situated in a ventral position in the head segment and is surrounded by a number of chitinous hooks.

The larvae are transparent in transmitted light, so one can see a number of the internal organs (fig. 15). The fat bodies (looking like long whitish sheets), the coiled intestine, and the yellowish Malpighian tubes are all visible. Embedded in the fat-bodies are the *gonads* (sex organs); these are large in the male but quite small in the female. Because of this size difference, it is possible to separate the sexes even at this early stage of development. Another obvious part of the internal anatomy are the two large lateral tracheae that run the full length of the larval body.

The three larval instars can be identified by microscopic examination of the mouthparts. These consist of a number of very dark chitinous plates (fig. 16). An H-shaped plate connects with a pair of hooks in front and a vertical pair of plates behind. At each moult, the mouthparts are shed together with the entire cuticle of the larva and can be found in the food. The new mouthparts are identical with the old, except for the pair of hooks. The number of teeth on these hooks changes with each moult (fig. 16). The first instar larva has one tooth on each hook, the second has two or three, and the third nine to twelve.

Just before pupation, the third instar larva normally leaves the food and crawls onto the sides of the culture bottle or onto paper towelling provided for this purpose (Chapter 8). The rather sluggish larva now turns its anterior spiracles inside out to form the pupal horns (fig. 17). The larva becomes motionless and begins to take on a more barrel-shaped pupal appearance. The last larval skin, which forms the case of the puparium, darkens and hardens. After about three-and-a-half hours the puparium is fully pigmented. In *D. melanogaster* the sex of a pupa can be determined easily by looking for the tarsal sex combs of the male (see p. 35). Once the combs have become pigmented they can be seen clearly, with the help of a binocular microscope, showing through the ventral surface of the

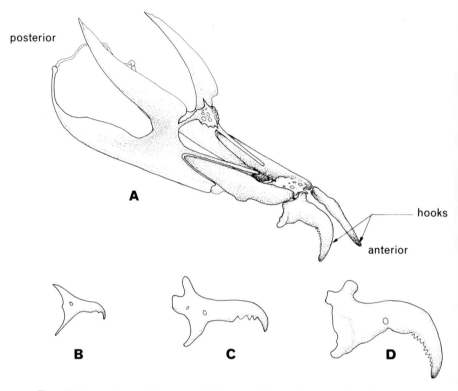

posterior

hooks

anterior

A

B   C   D

Fig. 16 Larval mouthparts of *Drosophila melanogaster*; **A** complete structure; **B** hook from first instar larva; **C** hook from second instar larva; **D** hook from third instar larva.

pupal case. In order that the pupa does not become dehydrated, this examination is best carried out on damp filter-paper.

The adult fly emerges from the puparium through the *operculum*. This lies on the dorsal surface of the case and, as the emerging adult exerts pressure, the seam around the operculum breaks. The newly emerged fly is unpigmented and is thin with unexpanded wings. Within a few hours the body becomes darker and more rounded and the wings expand.

Generally, the life-cycle of other species of *Drosophila* follows a similar path. In Figure 18 several kinds of eggs and puparia are shown to illustrate the variation found among the species. The length of time needed to complete the life-cycle can also vary between species, and the information given in the table on page 37 may be useful.

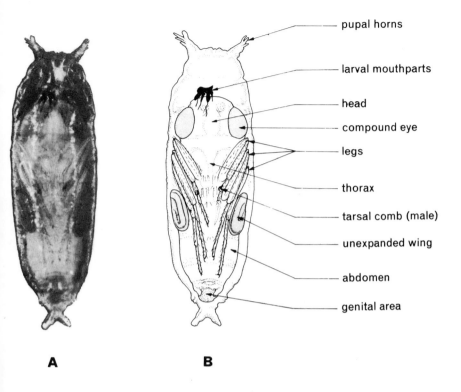

Fig. 17 Puparium of *Drosophila melanogaster*; **A** photograph of ventral aspect; **B** interpretative drawing of photograph.

A daily emergence rhythm has been observed in several species of *Drosophila*. The adults of *D. melanogaster*, when kept at a constant temperature and exposed to the normal sequence of day and night, emerge mostly between 0600 h and 1400 h. If cultures are kept for approximately fifteen generations under constant weak illumination, the emergence rhythm eventually disappears. However, a single brief exposure to darkness is sufficient to begin the rhythm again.

The length of adult life can vary considerably. Under ideal conditions when the flies are uncrowded and adequately fed, the average duration of adult life can be as long as 40 days in *D. melanogaster*. When the flies are crowded, even if supplied with fresh food daily, this figure may drop to 12 days. Under normal laboratory conditions, however, many workers

35

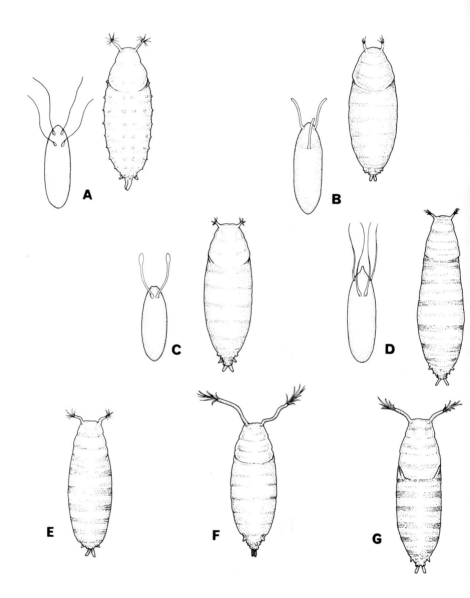

Fig. 18 Eggs and puparia of selected *Drosophila* species; **A** *D. busckii*;
B *D. phalerata*; **C** *D. subobscura*; **D** *D. littoralis*; **E** *D. funebris*;
F *D. immigrans*; **G** *D. hydei*.

	Days taken to complete life-cycle from egg to adult at:	
Drosophila species	18°C	25°C
D. ambigua	19–25	
D. ananassae		
D. andalusiaca		
D. busckii	21	12
D. cameraria	21	
D. deflexa	30–40	
D. fenestrarum		
D. funebris	23–28	13–15
D. helvetica		
D. hydei*	23–28	
D. immigrans	18–23	
D. kuntzei		
D. littoralis*		
D. melanogaster	18–20	8–10
D. obscura*	19–25	
D. phalerata	18–21	
D. subsilvestris	21	
D. simulans	16–18	9
D. subobscura	18–25	14–18
D. transversa		
D. tristis	21–26	

*The first few cultures do not normally take and this amount of time is not included in the table.
Where blanks appear, no information is available.

report that the adults die after an average period of only 6 or 7 days.

For many of the British species there is a scarcity of information about the length of adult life. Basden (1954*a*) reports that although *D. deflexa* usually die within a week or two of capture, one male lived for 115 days and one female for at least 150 days. Similarly, for *D. subsilvestris*, he records that of 71 adults emerging from apple-bait and kept in an outdoor insectary during the winter months, the last fly died after 167 days. Information of this sort for British *Drosophila* would be a most welcome addition to our biological knowledge.

When supplied with sufficient food, *D. melanogaster* females start laying eggs during the second day after emergence. Output then increases to a maximum on the fourth or fifth day and this is maintained for 3 to 10 days. The number of eggs produced then declines until it reaches zero shortly before the female dies. The total number of eggs laid varies widely. As many as 3168 have been recorded during a 70-day life-span.

Populations of *Drosophila* in the laboratory
A population of flies maintained in a bottle in the laboratory will vary in size according to certain environmental influences, the two most important being the amount of food and the amount of space available. An increase or decrease in the population is brought about by a change in the balance between the birth-rate and the death-rate. Crowded conditions and a lack of food produce a decrease in the egg-output of the females (a component of the birth-rate) and also an increase in adult death-rate. Similar effects can be found at all stages of the life-history.

Experiments have shown that the female will lay her eggs on the surface of the food in the bottle only when no other wet surface is available. For example, moist blotting-paper is preferred to standard laboratory food. The crucial factor appears to be the texture of the surface: a rough surface is usually preferred to a smooth one. Although they show this preference for certain surfaces, their *fecundity* (the number of eggs they lay) is not changed by the type of surface available.

Fecundity falls, however, when the flies are kept together in population bottles. After about 3 days from beginning the culture there is a progressive decrease in the number of eggs laid. In order to understand why this happens, it is important to realise that the flies feed mainly on the living yeast cells which are growing on the surface of the food (see Chapter 8). These yeast cells change in character as they grow and this, together with the growth of moulds on the food surface, produces the decline in egg-laying. A similar change in fecundity can be brought about experimentally by feeding different types of yeast to the female. In cultures which have been kept for some time there is often a shortage of yeasts on the surface of the medium and this may also contribute to the lowering of the egg-output.

Fecundity depends also on the history of the fly during the larval stage of its development, particularly on the amount and type of food available to it. Thus, each female is, to a certain extent, pre-conditioned so far as her future fecundity is concerned: small larvae produce small adults which, in turn, lay fewer eggs.

Not all eggs produce larvae, however, and there are two important factors which influence this. First, there is the kind of food (yeasts) available to the parent female both as a larva and as an adult. In this respect baker's yeast is better than brewer's yeast at the adult stage, though the reverse is true for the larval stage. When females are kept under favourable conditions, the number of *viable eggs* (eggs which can develop into larvae) increases to a maximum about the third day of her life, remains at this level for a few days, and then declines. This change in the viability of eggs with female age is also affected by the kind of food taken while she was a larva.

Second, many of the eggs get buried in the food as a result of the activity of the larvae around them. The number of eggs that disappear from the surface increases with the number and age of larvae present in the culture. In one experiment, 92% of uncovered eggs were viable, but covered eggs showed a very low viability of approximately 10%.

Once larvae begin to appear in the culture bottles a physical breakdown of the food medium takes place, brought about by their feeding activity. The small larvae remain in the surface layer but, later on, the workings of the larvae can extend to a depth of 5 mm and even deeper. For about 5 to 6 days (at 25° C) the medium is increasingly broken up by their tunnels; this has been called the *aerated* phase. On about the sixth day, the tunnels collapse and the food becomes pulpy and waterlogged. This change is of great importance in altering the nature of the yeasts present. In these new conditions, the respiration of the yeasts passes from an *aerobic phase* to an *anaerobic phase*, and this leads to fermentation.

A reduction in food yields smaller larvae which, in turn, give smaller pupae. Every larva, however, needs a certain minimum quantity of food in order to pupate. Below a certain critical size pupation does not take place. The size of the pupae has a marked effect upon their survival: smaller pupae show a lower rate of survival. Larvae that have been inadequately fed may be able to pupate, but the pupae thus formed may fail to develop into adults.

It has already been pointed out that food shortage and crowding affect adult survival. The conditions in the culture bottle—temperature, humidity and so on— also play a part. In most cases adults are much less able to withstand extremes of the physical environment than are the other stages in the life-history.

The *Drosophila* culture bottle is a complex ecological situation. Any examination of this system must take into account all the life-stages present. Within each stage and between each stage there is competition for the necessities of life. This is known as *intraspecific competition* because only one species is involved. In addition, there are interactions between the *Drosophila*, the yeasts and moulds, and the physical and chemical environment.

The situation can be made even more complex if two species of *Drosophila* are introduced into the same culture bottle (*interspecific competition*). Both species are now competing for food and living space.

It is unlikely that both would be equally successful and, therefore, one species would eventually cease to exist. When *D. melanogaster* and *D. simulans*—two closely related species—are cultured together at 25°C, it is usual for *D. simulans* to become less abundant until it eventually disappears altogether after about 100 days. At 15°C, *D. melanogaster* is the unsuccessful species, however. Similarly, the competitive ability of *D. simulans* can be improved if its numbers are high, in comparison with *D. melanogaster*, in the founding population. This kind of result suggests that, in the wild, a competitively weaker species might prevent a stronger one from becoming established in an area simply by being there in large enough numbers.

Another example of a competitive situation is that found between *D. melanogaster* and *D. funebris*. If both these species are maintained in a population cage (see Chapter 8) and no new food is added, *D. funebris* is the successful species. However, if fresh food is added at intervals, both species are able to survive. The survival of both types is the result of fluctuations in the environment which favour first one species and then the other. Most of the competition occurs at the larval stage. The fresh, newly yeasted medium is more favourable to the development of *D. melanogaster* larvae. As the food ages and other micro-organisms appear, it becomes more suitable for the growth of *D. funebris* larvae. Thus, because those parts of the environment used by each species do not overlap completely, both can exist side by side. Such mixed populations have been maintained in the laboratory for periods of up to two years without the loss of either species.

3 FIELD ECOLOGY

Studies on the natural ecology of fruit flies have been carried out mainly in the United States and South America. In Great Britain, ecological information is relatively hard to find and more field work needs to be conducted.

Distribution of *Drosophila* in space
Much work has been concerned with the distribution of *Drosophila* in space and the factors influencing these distributions. For example, in the table below are some data collected by Dobzhansky and Epling (1944) in California, showing the numbers of individuals of *D. pseudoobscura* coming to traps arranged in a line at 20-m intervals. Traps A to G stood in a ravine densely shaded by coniferous trees and along which flowed a small stream; traps H to K were at the edge of a forest among large oaks; traps L to O extended into a grassy meadow. Most flies were collected in traps H to K. With *D. pseudoobscura* this kind of result was obtained consistently in several areas. Traps exposed near oak trees collected large numbers of this species.

	trap	A	B	C	D	E	F	G	H	I	J	K	L	M	N	O
June 5		3	1	9	0	7	8	8	15	11	29	80	10	7	4	1
June 6		6	6	7	1	4	5	3	5	2	14	29	5	1	1	0
June 8		7	5	10	4	6	4	10	11	19	33	43	8	3	3	2
June 10		2	7	0	0	3	8	9	4	11	19	52	7	3	2	11
June 11		5	1	1	1	5	2	5	4	3	17	23	8	6	1	2

Dobzhansky and Pavan (1950) counted the numbers of various species of *Drosophila* in a forest in Brazil. Baits were set at 10-m intervals along two transects, each 200 m long, which crossed in the middle at right-angles. The baits attracted flies from the surrounding area, which were collected throughout the day by sweeping over the baits with a net. Three sets of results for three separate days are shown in Figure 19. The

28 December, 1948   5 February, 1949

11 June, 1949

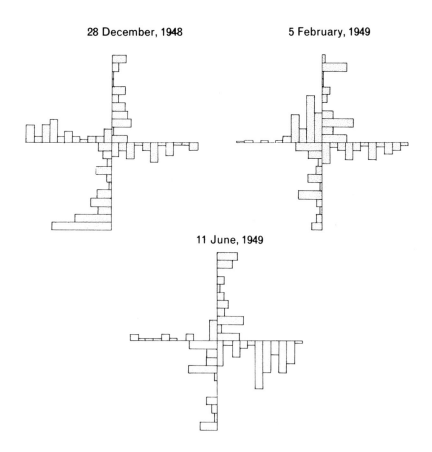

Fig. 19 The distribution of *Drosophila* species in a Brazilian forest. Traps were arranged in the form of a cross. Flies coming to the traps were counted on three separate occasions and the heights of the shaded columns give an indication of the numbers of flies caught at each trap.

lines at right-angles represent the transects, and the heights of the columns along these lines indicate the numbers of flies captured at each site. The distributions are extremely patchy and the pattern varies from month to month. The pattern also varies for different species, which are attracted to different kinds of food. Moisture and light, both of which may vary in a very irregular way depending upon the amount and kind of vegetation, are also influential factors.

Similar data have been collected in this country (Shorrocks, 1970). At four sites, within what seemed to be a uniform deciduous woodland, all within 100 m of each other, differences in species numbers collected were noted. For both *D. phalerata* and *D. obscura* the majority of the individuals (80% and 64% respectively) were caught at one site. For *D. subobscura*, 66% of the total catch was obtained from just two of the sites. These results show that a collection of fruit flies made at only one site in an area could give a misleading impression of the species present.

An interesting distribution pattern has been found along the Tyne Valley, Northumberland (Shorrocks, 1969). Fourteen collection sites were visited and the proportions of the three commonest species are shown in Figure 20. Only *D. subobscura* penetrates to the centre of the city.

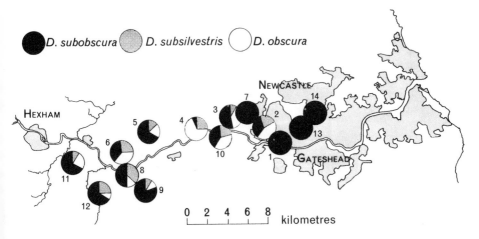

Fig. 20 The distribution of *D. subobscura, D. obscura,* and *D. subsilvestris* along the Tyne Valley, Northumberland.

D. obscura is absent from the city centre, but reaches its highest frequencies in the surrounding areas (sites 2, 3, 4, and 10). West of site 4 there is again a decrease. *D. subsilvestris* makes up a small but constant percentage of all flies collected along the whole length of the river. It is absent only from the city and domestic habitats (sites 1, 7, 13, and 14).

Fruit flies are also distributed vertically through space. By placing traps at various heights in a woodland this distribution can be investigated. Basden (1953) placed apple traps at different heights in 4 deciduous trees. The flies coming to these traps were collected over a 3-month period. When the trees had no leaves, most of the flies came to the traps on the ground.

However, when the three mature trees were in leaf, the majority of the flies were taken in the upper traps (10-15 m). In the case of the one younger tree (in a narrow belt of trees of a uniform height of about 15 m) most of the flies occurred in the ground-level traps. Traps placed in a 20-m conifer showed that, here, the flies remained in the sheltered crown of the tree long after they had forsaken the bare crowns of the deciduous trees.

To understand these distribution patterns we must examine data on feeding and breeding sites. In Chapter 7, detailed information about 22 British species is given. Here we will outline the feeding and breeding sites of *Drosophila* in more general terms.

In the wild, both adult flies and larvae feed on yeasts and bacteria which occur in fermenting substances rich in carbohydrates. It is possible to recognize several more or less distinct kinds of feeding and breeding habits: general scavengers, fruit-feeders, sap-feeders, fungus-feeders, and species which use flowers. Although some species are restricted to only one of these categories, many are not. *D. subobscura*, although regarded as a sap-feeder along with others belonging to the obscura group, will also use many kinds of fruit. *D. immigrans* and *D. funebris*, both very common, widespread species regarded as fruit-feeders and general scavengers respectively, have been found in fungi, especially if the fungus' fruiting-body has begun to decay. The specialization of diet appears to be carried to even greater extremes in some species. *D. phalerata* and *D. transversa* both feed and breed on fungi. However, what little information is available indicates that while *D. phalerata* has quite a wide taste, *D. transversa* is much more selective.

An example of even greater ecological specialization has been discovered by Heed and Kircher (1965) working in the United States. *D. pachea* is a species found only in the Sonoran Desert (south-western United States and Mexico). It breeds only in the stems of the senita cactus and is incapable of reproducing in the laboratory unless a piece of this cactus is added to the medium. The substance necessary for its growth and reproduction is a sterol called *schottenol* which is present in this particular cactus. Several alkaloids have also been isolated from it and these appear to inhibit or kill the larvae of other *Drosophila* species inhabiting the Sonoran Desert. *D. pachea*, however, is tolerant to these alkaloids.

An investigation carried out in the eastern United States by Carson and Stalker (1951) on *D. robusta* suggests that for this species at least there is a difference between feeding and breeding sites. It appears to feed and breed on the sap exuded from various trees. However, evidence indicates that when sap is suitable for egg-laying it attracts very few feeding flies, and *vice-versa*. For example, the sap coming from one particular oak tree was highly attractive to *D. robusta* adults during the month of June, although no eggs were then being laid there. At the same time, sap on an elm tree only 25 m away was being used for oviposition. In July, when the oak tree became suitable for oviposition, its attractiveness as a feeding site seemed reduced. The authors attribute their observations to changes in the micro-organisms inhabiting the sap. The succession of events that occurs from the time a sap-flow begins to the time it finally dries up are obviously important to the *Drosophila* species using them. To say that certain *Drosophila* feed and breed on sap-flows is much too simple: detailed ecological investigation is needed.

A similar succession of events will probably occur in the fruiting body of a fungus, so that for fungus-feeders also we need more detailed information about feeding and breeding sites.

Distribution of *Drosophila* in time

Pavan and his co-workers (1950) recorded the number of *Drosophila* species visiting traps between the hours of 0500 h and 1800 h in an open wood in Brazil. All of the 10 species recorded had a similar pattern of activity. Flies visited the traps chiefly from 0600 h to 0800 h and from 1630 h to 1800 h; there was no activity during the night. The pattern of activity would seem to be an adaptation which prevents the flies from moving about when conditions are unfavourable. The traps were visited by the flies mainly when the temperature was low and the humidity high. For example, within a tropical rain forest there is no noticeable drop in activity during the middle of the day, but at the edge of a similar forest the activity is lower during periods of sunshine.

A similar study of the daily activity rhythm of *D. subobscura* and *D. obscura* was carried out by Dyson-Hudson (1956) in Britain. Paper cups containing fermenting banana were hung from trees approximately 1 m from the ground and at an angle of about 45°. Collections of flies were made at invervals of 1½ to 2 h throughout the day, from before dawn until after dark. Temperature, humidity, and light were recorded at the time of each collection. The results showed that, during the summer in woodlands with a dense leaf canopy, the activity of *D. subobscura* begins at dawn, continues throughout the day (with some indication of a morning peak), and rises to a marked peak in the evening. In open meadows, there is a marked morning and evening peak of activity but very few flies are caught during the middle of the day. In open woodlands, *D. subobscura* shows an intermediate pattern of activity (fig. 21).

Dyson-Hudson suggests that this activity pattern is determined by two factors in the environment: temperature and light. In the morning the light intensity is suitable but the temperature is not; this accounts for the small morning peak. By the time the temperature is right, the light is too

45

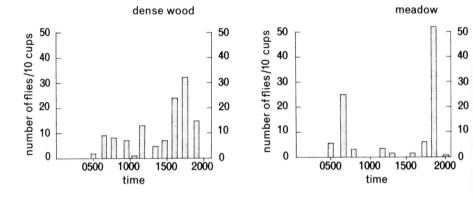

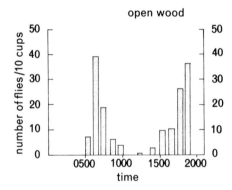

Fig. 21 The daily rhythm of activity of *Drosophila subobscura* in dense wood, open wood, and meadow.

intense and, therefore, little activity occurs during the middle of the day, except in dense woodlands. By evening both factors are ideal for activity and consequently a peak is observed.

Collections made during early spring or late autumn show that, at these times, activity and temperature follow each other very closely. There is little activity below 10°C. Although activity is evident above this temperature, large numbers of flies are not caught at temperatures below 15°C. The difference between "summer" and "winter" activity is determined by environmental conditions and not by the time of year, because on warm days in April and May a "summer" activity pattern is observed.

With *D. obscura* the pattern of activity throughout the day is different. Since this species is restricted more to woodland habitats the observations cover a limited range of environments; also, data was available only for "summer" months. No constant pattern was found because there was great variation in activity on different days within the same woodland. All that can be said about this species is that activity begins at dawn, fluctuates throughout the day, and stops just after dusk.

Many workers have collected fruit flies at regular intervals throughout the year to look for seasonal activity patterns. Patterson (1943) made a study of seasonal variation near Austin, Texas. He collected 31 species, although several he collected in such small numbers that it was impossible to deduce any pattern of variation at all. The flies were most abundant in autumn and spring. Some species exhibited peaks at both these times and others at only one of them. For example, many of the fungus-feeding species showed an increase in numbers in autumn. Chapter 7 gives details of the seasonal activity found in many British species.

Dobzhansky and Epling (1944) point out that many of these seasonal cycles are flexible and depend upon the climatic conditions which prevail. Certainly this has been found to be the case in the North American *D. pseudoobscura*. In warmer climates, this species has a spring maximum in March and April; in cooler climates, including higher altitudes in the same general area, the peak occurs in June and July. Similarly, in warmer localities, *D. pseudoobscura* may be present all winter and nearly disappear in the heat of summer; in colder localities this pattern may be reversed.

Overwintering

During the winter months in Britain most species of fruit flies are captured only rarely. The manner in which these flies pass the unfavourable winter months is a problem that has not been examined in great detail, although it is obviously a matter of great importance to the flies' survival. Field observations on some Japanese species have shown that adults are quite capable of surviving the winter out of doors by forming colonies in the ground, just beneath the snow.

Basden (1954*b*) has shown that the larvae of *D. deflexa* pass into a state of arrested development, called *diapause*, from autumn until spring. This happens whether they are kept outdoors or in the laboratory at a constant temperature of 18°C. In the North American species, *D. persimilis*, there is some suggestion that at high altitudes diapause occurs in the pupal stage. Carson and Stalker (1948) working in Missouri, in the United States, found that females of *D. robusta* captured in the wild from August onwards reproduced much more slowly than those collected in spring or early summer. This reproductive diapause occurred despite the fact that temperatures were still favourable in the wild. Upon examination, Carson and Stalker found that the autumn population of females consisted mainly of virgin flies with underdeveloped ovaries and large amounts of body fat. They came to the conclusion that *D. robusta* over-winters as an adult and that this period of hibernation is preceded by a change in body processes from egg-production to the laying down of fat.

Predators and parasites
Information on the natural predators of *Drosophila* is almost completely lacking although, no doubt, a whole range of animals include *Drosophila* in their diet. The common yellow dung fly (*Scatophaga stercorarium*) can be seen often in the vicinity of fermenting substances waiting for the flies that will be attracted to this food. These same flies also wait by *Drosophila* traps and the author has observed them preying upon the visiting fruit flies.

Two species of parasitic wasps, *Pseudeucoila bochei* (Hymenoptera: Cynipidae) and *Phaenocarpa tabida* (Hymenoptera: Braconidae) are known to use *Drosophila* species as hosts. Both are small (2 mm long) and dark-brown/black in colour and can be reared easily from *Drosophila* traps left in the wild for some time. The female *Pseudeucoila* lays an egg inside a *Drosophila* larva. The egg hatches within about 48 hours. The young wasp develops inside the *Drosophila* larva, which continues its development and finally pupates. The parasitic wasp larva completes its development in the host puparium and emerges as an adult after about three weeks. The life history of *Phaenocarpa* is very similar except that the wasp larva may go into diapause, in which case the development time is extended by several weeks.

Experiments with *Pseudeucoila* have shown that the one wasp to one *Drosophila* larva relationship is caused by two factors. There is discrimination between healthy and already parasitized hosts on the part of the egg-laying female. This discrimination is not absolute, however, and depends strongly upon the availability of healthy hosts. If two eggs are laid within one *Drosophila* larva the first wasp-larva to hatch destroys the other so that only one adult wasp emerges from the puparium.

Data collected on the Continent would suggest that *Phaenocarpa* is not so common as *Pseudeucoila*. This latter species appears to parasitize in the wild mainly *D. melanogaster*, but *D. subobscura*, *D. hydei*, *D. funebris*, and *D. littoralis* are also utilized. Both the developmental time and the size of the wasps are influenced by the species of *Drosophila* parasitized. Adults from the larger species of *Drosophila* (*D. littoralis* and *D. funebris*) are often larger.

4 GENETICS

Drosophila is an ideal animal for studying the laws of inheritance. These laws, initially formulated by Gregor Mendel in 1865 represented a tremendous step forward from previous thought on the subject. Until that time people had favoured the idea of *blending inheritance*, the idea that a substance from each parent formed an inseparable mixture in their offspring and determined the inherited characters.

Mendel suggested that the material determining the characteristics of the offspring consisted of a number of separate units transmitted to the progeny in the sperm and egg of the parents. These units, or *genes* as they are now called, do not blend with one another and can be passed on, therefore, to future generations intact. Genes are carried in the *chromosomes* (see Chapter 5). Since the chromosomes are typically in pairs, each gene is represented twice in a cell, once in one chromosome and once again at the same *locus* (position) in the other member of the pair. Genes at the same locus are known as *allelomorphs* or *alleles*.

From time to time genes are *not* passed on to the next generation in exactly the same form as those of the parents. These *mutated* (changed) genes affect the characteristics of the offspring, producing what are known as *mutant forms* of the species. The mutant individuals may be able to transmit their mutant genes to their progeny so that populations of mutant flies can be maintained. Most of the examples used in this chapter concern *D. melanogaster*, in which a large number of such mutants are available for study.

Mendel's first law—the law of segregation
If an individual of *D. melanogaster* from a normal (often called wild-type) strain of flies is mated with an individual from a strain having the mutant character known as *ebony* (see Appendix), the offspring are all normal in appearance. However, on mating two of these "normal" looking flies, both normal and ebony individuals appear in their progeny. The two types are found in a specific ratio: there are always approximately three times as many normal as ebony individuals. This has a simple genetic explanation.

In the formation of the *gametes* (eggs and sperm) the pairs of chromosomes separate and one member of each pair passes into a different gamete.
Thus, the number of chromosomes in the gametes is half that in the adult
body cells. (Geneticists usually talk of gametes as having a *haploid* set of
chromosomes and body cells having a *diploid* set.) Each gene, therefore,
is represented only once in each gamete until fertilization, when the fusion
of a male and female gamete restores the diploid condition.

Ebony individuals possess mutant alleles that are responsible for the
black coloration of the body, but normal individuals possess normal (or
wild-type) alleles at the corresponding locus. Consequently, a normal

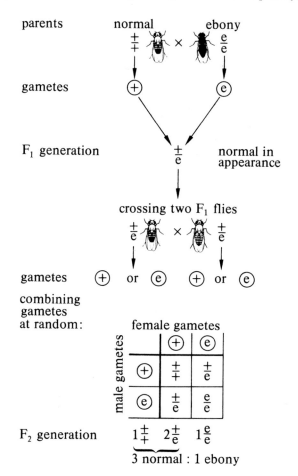

parent passes on normal alleles to the offspring while the ebony parent passes on a mutant allele. Thus, all the offspring (often called the *first filial* or F_l *generation*) possess both a normal and a mutant allele for body coloration. These F_1 individuals can pass on to their progeny (the F_2 *generation*) both sorts of alleles, and the random mating of F_1 flies produces a 3:1 ratio of normal:ebony flies. The diagram on p. 50 illustrates this. The normal allele is represented by the symbol +. The line separating the two normal symbols represents the pair of identical chromosomes (*homologous chromosomes*) upon which the alleles are located. The symbol e is used to denote the ebony allele.

Individuals possessing similar allelomorphs are called *homozygotes* (+/+ or e/e). Individuals possessing different allelomorphs are known as *heterozygotes* (+/e). In the example just given, the ebony allele does not have any effect in the heterozygote and, for this reason, it is known as a *recessive character*. The normal appearance is said to be *dominant*. Because ebony is recessive it is represented symbolically by a small letter. Dominant mutants are given a capital letter. In the above example, it can be seen that of the flies with a normal *phenotype* (appearance), only one-third are homozygotes and will, therefore, breed true. Two-thirds are heterozygotes possessing a different *genotype* (combination of genes). Thus, flies with similar phenotypes can possess different genotypes. This cross also demonstrates that genes can be concealed in a particular generation, but remain unchanged, and be passed on to future generations where they can affect the nature of the phenotype once again. An experiment such as this can quite easily be carried out in the laboratory (see Investigations).

Mendel's second law—independent assortment

Let us now consider what happens when we cross a normal individual with an individual possessing two mutant characters determined by genes situated on different chromosomes. We shall use the mutant ebony, which is recessive and is situated on chromosome 3, and the mutant *vestigial wings* (vg), which is also recessive but is situated on chromosome 2. The diagram on p. 52 sets out the pedigree and uses the notation already introduced. Notice that since the pairs of alleles involved are on different chromosomes, the lines separating the symbols are *not* continuous.

In the F_1 generation the flies are all heterozygotes but have a normal appearance because both mutants are recessive and can, therefore, exert their influence only in a homozygous form. The F_1 individuals can produce four types of gametes which, when united at random, will give nine genotypes. But, because of the dominance of the normal allelomorph, only four phenotypes appear, in the ratio 9:3:3:1.

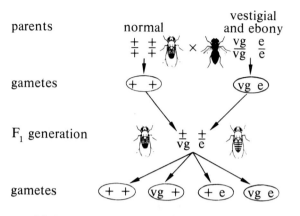

parents   normal   vestigial and ebony

$$\frac{+}{+} \; \frac{+}{+} \; \times \; \frac{vg}{vg} \; \frac{e}{e}$$

gametes $(+ \; +)$ $(vg \; e)$

F_1 generation $\frac{+}{vg} \; \frac{+}{e}$

gametes $(+ \; +)$ $(vg \; +)$ $(+ \; e)$ $(vg \; e)$

combining gametes at random:

female gametes

male gametes	$(+ \; +)$	$(vg \; +)$	$(+ \; e)$	$(vg \; e)$
$(+ \; +)$	$\frac{+}{+} \; \frac{+}{+}$ normal	$\frac{+}{vg} \; \frac{+}{+}$ normal	$\frac{+}{+} \; \frac{+}{e}$ normal	$\frac{+}{vg} \; \frac{+}{e}$ normal
$(vg \; +)$	$\frac{vg}{+} \; \frac{+}{+}$ normal	$\frac{vg}{vg} \; \frac{+}{+}$ vestigial	$\frac{vg}{+} \; \frac{+}{e}$ normal	$\frac{vg}{vg} \; \frac{+}{e}$ vestigial
$(+ \; e)$	$\frac{+}{+} \; \frac{e}{+}$ normal	$\frac{+}{vg} \; \frac{e}{+}$ normal	$\frac{+}{+} \; \frac{e}{e}$ ebony	$\frac{+}{vg} \; \frac{e}{e}$ ebony
$(vg \; e)$	$\frac{vg}{+} \; \frac{e}{+}$ normal	$\frac{vg}{vg} \; \frac{e}{+}$ vestigial	$\frac{vg}{+} \; \frac{e}{e}$ ebony	$\frac{vg}{vg} \; \frac{e}{e}$ vestigial and ebony

F_2 generation 9 normal : 3 vestigial : 3 ebony : 1 vestigial and ebony

The pedigree illustrates that when pairs of alleles are sited on different chromosomes they are distributed to the gametes independently. Once again, the pattern of inheritance is easily confirmed in the laboratory using the mutants named (see Investigations).

Students are often puzzled by the fact that frequently an exact 3:1 or 9:3:3:1 ratio is not obtained in these laboratory experiments. Apart from deviations caused by chance (which may be quite large with small numbers of offspring), it must be remembered that the ratios refer to newly fertilized eggs. During the development of the fly to the adult stage when these characters are visible, mortality occurs which is often more severe among mutants than among normal individuals. Consequently, a deficiency of mutants, particularly of double homozygotes (vg/vg e/e), can sometimes be detected.

Linkage
Genes are carried on the chromosomes and so their distribution in the gametes is controlled by the distribution of the chromosomes. If two gene

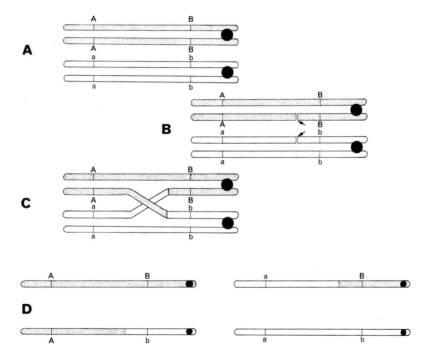

Fig. 22 The sequence of events in crossing over; **A** pairing of homologous chromosomes (four-chromatid stage); **B** a break occurs in two of the chromatids; **C** the broken ends of different chromatids rejoin, resulting in a cross-over; **D** the four chromatids (chromosomes) that result from crossing over.

loci are close together on the same chromosome, they will be carried together into the same gamete and will not assort independently. During the cell-divisions which result in the halving of the chromosome number and the production of gametes, genetic material may be exchanged between the chromosomes of each homologous pair. Each chromosome comes to lie next to its partner so that each gene is closely associated with its allelomorph. The chromosomes now divide longitudinally so that each consists of two identical strands called *chromatids*. This stage is shown diagrammatically in Figure 22A. Sometimes, a break occurs in one of these chromatids and this appears to bring about a similar event at the same place in one of the other three chromatids (fig. 22B). On rejoining, the broken ends may cross over so that new combinations of genes result (fig. 22C). Two cell-divisions separate the four chromatids so that four gametes, each containing one of the four chromatids, are formed (fig. 22D). Each gamete will have its own genotype; AB and ab are like the parent chromosomes and are called *parental classes*. Ab and aB are new combinations and are called *recombinant* or *cross-over classes*. This whole process is known quite simply as *crossing over* or *recombination*.

Let us consider the two individuals (A/A B/B and a/a b/b) with the loci A and B not linked; that is, they are on separate chromosomes. On mating, these will give an F_1 consisting of double heterozygotes (A/a B/b). If one of these F_1 individuals is paired with a fly having the genotype of the recessive parent (a/a b/b), four classes of F_2 offspring are obtained. The next diagram illustrates this. However, if the genes were very closely linked

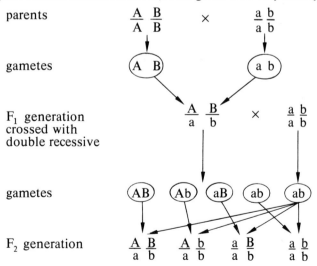

54

and there was no crossing over, the pedigree would have the appearance shown in the next diagram and the only classes produced in the F_2 would be A/a B/b and a/a b/b, in equal numbers.

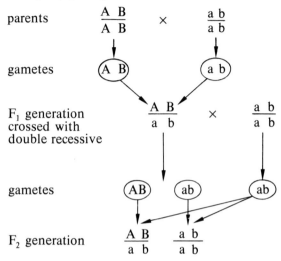

parents $\dfrac{A\ B}{A\ B}$ × $\dfrac{a\ b}{a\ b}$

gametes (A B) (a b)

F_1 generation crossed with double recessive $\dfrac{A\ B}{a\ b}$ × $\dfrac{a\ b}{a\ b}$

gametes (AB) (ab) (ab)

F_2 generation $\dfrac{A\ B}{a\ b}$ $\dfrac{a\ b}{a\ b}$

If some crossing over occurred a small proportion of the gametes Ab and aB would be found. These gametes would be in equal numbers because one of each kind results from a single cross-over. Consequently, among the offspring there will be all four possible genotypes, but the majority would consist of A/a B/b and a/a b/b individuals in equal numbers, and the minority of A/a b/b and a/a B/b, again in equal numbers. For example, we might get the following:

$$36\,\dfrac{A\ B}{a\ b} \qquad 13\,\dfrac{A\ b}{a\ b} \qquad 12\,\dfrac{a\ B}{a\ b} \qquad 39\,\dfrac{a\ b}{a\ b}$$

From such data it is possible to calculate the *cross-over value*—that is, the percentage of individuals that represent recombinant classes. With the data just quoted this value would be:

$$\frac{13+12}{36+13+12+39} = \frac{25}{100} = 25\%$$

The cross-over value between two genes gives some measure of their distance apart on the chromosome. The farther apart the genes, the greater will be the chance of a cross-over occurring between them and the larger the cross-over value. Thus, chromosomes can be mapped in terms of cross-over values obtained from genetic experiments. Under this system,

55

*one unit of map distance between linked genes is the space within which 1%
crossing over takes place.*

Drosophila geneticists have found that an efficient way to obtain good
recombination data is to use *three-point test-crosses* involving three genes
situated within a relatively short segment of a chromosome. An example
of such a cross is given here, where three recessive genes are designated by
the letters a, b, and c. The genotypes occurring in the F_2 can be classified
and their numbers recorded.

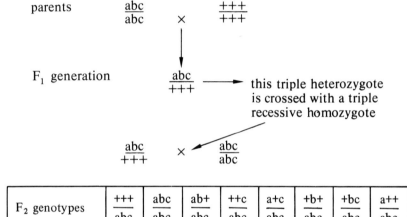

F_2 genotypes	$\dfrac{+++}{abc}$	$\dfrac{abc}{abc}$	$\dfrac{ab+}{abc}$	$\dfrac{++c}{abc}$	$\dfrac{a+c}{abc}$	$\dfrac{+b+}{abc}$	$\dfrac{+bc}{abc}$	$\dfrac{a++}{abc}$
no. of offspring	235	270	4	7	62	60	40	48

The triple heterozygote will form eight kinds of gametes (top line of
genotypes above) which can be traced easily to the progeny because the
gametes contributed by the triple recessive homozygote contain only
recessive genes.

In analysing the progeny from a three-point test-cross, the first step
is to identify the parental (non-recombinant) gametes from the hybrid
parent (abc/+ + +). In the present example these are the classes + + +/
abc (235) and abc/abc (270). These classes include by far the largest
numbers of individuals. The sequence of genes (abc) has been denoted in
a purely arbitrary fashion and may not represent the true gene-order at all.
We can determine the correct gene-order, however, by comparing the
parental combination of genes with that of the double cross-over classes.
Since double crossing over is the simultaneous occurrence of two relatively
improbable events, the double cross-over classes are the least frequent

type (classes ab+/abc and ++c/abc). The allelic pair +/c has been exchanged in order to make double cross-over chromosomes from parental chromosomes. That allelic pair must be situated between the other two. Thus, the actual gene sequence is acb.

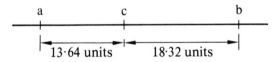

The cross-over value between a and c is:

$$\frac{40+48+4+7}{total} \times 100$$

$$= \frac{99}{726} \times 100 = 13\cdot64\%$$

The cross-over value between c and b is:

$$\frac{62+60+4+7}{total} \times 100$$

$$= \frac{133}{726} \times 100 = 18\cdot32\%$$

We can now draw the chromosome map for these three gene loci, indicating their relative distance apart in terms of map units or percentage crossing over between two loci.

```
     a                 c              b
 ────┼─────────────────┼──────────────┼────
    ├───────────────►├◄─────────────────►┤
      13·64 units        18·32 units
```

Once a portion of a chromosome map is established, it provides the geneticist with a table of values which indicate the likelihood or probability of obtaining a cross-over between any two gene loci. For example, the probability of obtaining a gamete with a cross-over between c and b above is 18·32% or 0·1832. If, in the portion of chromosome just mapped, crossing over in the region between a and c is independent of that in the region between c and b, the probability of simultaneous crossing over in the two regions can be found by multiplication of the two independent probabilities. In other words, if crossing over in the two regions is

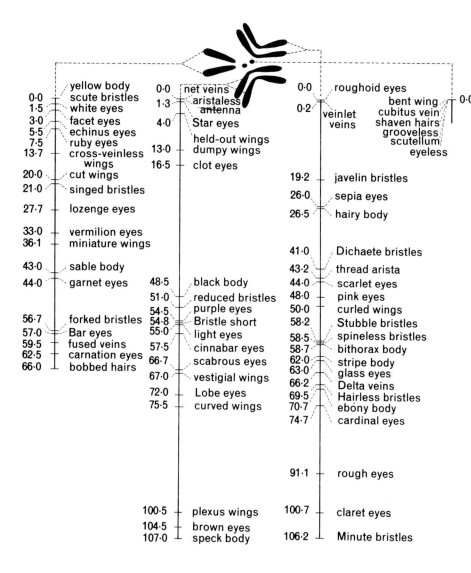

0·0	yellow body	0·0	net veins	0·0	roughoid eyes
0·0	scute bristles	1·3	aristaless		bent wing
1·5	white eyes		antenna	0·2	cubitus vein
3·0	facet eyes	4·0	Star eyes		veinlet
5·5	echinus eyes				veins
7·5	ruby eyes		held-out wings		shaven hairs
13·7	cross-veinless	13·0	dumpy wings		grooveless
	wings	16·5	clot eyes		scutellum
20·0	cut wings				eyeless

0·0 (top right)

Fig. 23 A genetic or linkage map of the four chromosomes of *Drosophila melanogaster* showing the positions of some of the more important genes. Genes labelled with an initial capital letter are dominant.

Chromosome 1 (left):
0·0 yellow body
0·0 scute bristles
1·5 white eyes
3·0 facet eyes
5·5 echinus eyes
7·5 ruby eyes
13·7 cross-veinless wings
20·0 cut wings
21·0 singed bristles
27·7 lozenge eyes
33·0 vermilion eyes
36·1 miniature wings
43·0 sable body
44·0 garnet eyes
56·7 forked bristles
57·0 Bar eyes
59·5 fused veins
62·5 carnation eyes
66·0 bobbed hairs

Chromosome 2 (middle):
0·0 net veins
1·3 aristaless
antenna
4·0 Star eyes
held-out wings
13·0 dumpy wings
16·5 clot eyes
48·5 black body
51·0 reduced bristles
54·5 purple eyes
54·8 Bristle short
55·0 light eyes
57·5 cinnabar eyes
66·7 scabrous eyes
67·0 vestigial wings
72·0 Lobe eyes
75·5 curved wings
100·5 plexus wings
104·5 brown eyes
107·0 speck body

Chromosome 3 (right):
0·0 roughoid eyes
bent wing
0·2 cubitus vein
veinlet veins
shaven hairs
grooveless
scutellum
eyeless
19·2 javelin bristles
26·0 sepia eyes
26·5 hairy body
41·0 Dichaete bristles
43·2 thread arista
44·0 scarlet eyes
48·0 pink eyes
50·0 curled wings
58·2 Stubble bristles
58·5 spineless bristles
58·7 bithorax body
62·0 stripe body
63·0 glass eyes
66·2 Delta veins
69·5 Hairless bristles
70·7 ebony body
74·7 cardinal eyes
91·1 rough eyes
100·7 claret eyes
106·2 Minute bristles

58

independent, $0.183 \times 0.136 = 0.025 = 2.5\%$ double cross-overs might be expected. In fact, only $11/726 = 1.5\%$ actually occurred. This seems to suggest that once crossing over occurs, the probability of another cross-over in an adjacent region is reduced. This phenomenon is called *interference*. Strengths of interference are measured as *coefficients of coincidence*:

$$= \frac{\text{actual frequency of double cross-overs}}{\text{expected frequency of double cross-overs}}$$

For the above data this equals:

$$\frac{1.5}{2.5} = 0.6$$

The chromosome maps of *D. melanogaster* are shown in Figure 23.

Sex-linkage
Since the sex-chromosomes determine the sex of the fly, genes which are situated on the so-called X-chromosomes are likely to show a form of inheritance linked with maleness and femaleness. The two crosses following illustrate some of the peculiarities of sex-linked inheritance. Male

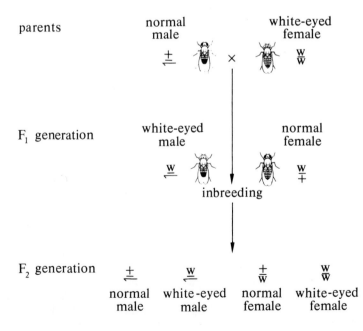

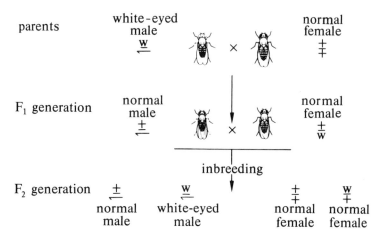

parents — white-eyed male $\underset{\longleftarrow}{w}$ × normal female $\dfrac{+}{+}$

F$_1$ generation — normal male $\underset{\longleftarrow}{+}$ × normal female $\dfrac{+}{w}$

inbreeding

F$_2$ generation

$\underset{\longleftarrow}{+}$	$\underset{\longleftarrow}{w}$	$\dfrac{+}{+}$	$\dfrac{w}{+}$
normal male	white-eyed male	normal female	normal female

Drosophila have only one X-chromosome; the other member of the pair is the Y-chromosome, represented by the symbol ←. The sex-linked mutant white-eyed (w) is used.

The female parent, with two X-chromosomes (and no Y-chromosome), seems to transmit her mutant character only to her male progeny, a feature typical of sex-linked inheritance. In the F$_2$ generation, both males and females display normal and mutant characters in equal proportions. Notice in the latter cross how the appearance of white eyes in the F$_2$ generation is linked with maleness. Notice also how the results of the two reciprocal crosses have differed, depending on whether the mutant allele is introduced through the male or female parent.

Genes in populations

Until now we have considered only genes in families or pedigrees; this section deals with genes in populations of fruit flies. Most of the mutant genes that have been observed and used in genetical experiments with *Drosophila* convey some disadvantage on their carriers. In addition, the majority of these are recessive in character. Because of these two facts, although mutant phenotypes (homozygotes) are rarely found in the wild, a large number of recessive mutations lie concealed in heterozygous form. It is a fairly easy matter to collect flies from the wild and, through a short programme of inbreeding, uncover some of this concealed genetic variability. Most female flies caught in the wild have already been inseminated. She or the male may have carried a rare recessive mutant gene (m) concealed in heterozygous form, but it is unlikely that both did. Consequently, the presumed mating before capture can be depicted as:

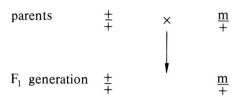

In the F_1 generation produced by the captured female, one half are expected to be heterozygous for the concealed mutation. If these F_1 generation flies are allowed to interbreed, four possible matings can occur

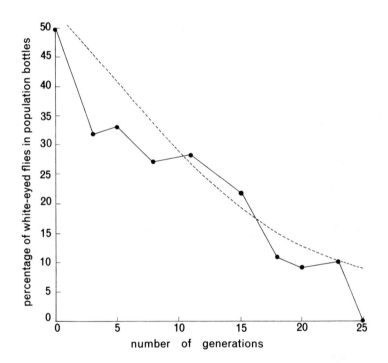

Fig. 24 The elimination by natural selection of white-eyed flies from a laboratory population. The solid line shows the elimination from an actual population. The broken line shows the expected decrease in the white-eyed population assuming that white-eyed males are 75% less successful at mating than are normal flies.

to produce the following F_2 progeny:

$+/+$ males $\times$ $+/+$ females will give all $+/+$

$\left.\begin{array}{l} +/+ \text{ males} \times +/m \text{ females} \\ +/m \text{ males} \times +/+ \text{ females} \end{array}\right\}$ will give $1\ +/+$:$1\ +/m$

$+/m$ males $\times$ $+/m$ females will give $1\ +/+$:$2\ +/m$:$1\ m/m$

Therefore, if one of the original parents carried a recessive mutant in heterozygous condition, then $\frac{1}{16}$ of the F_2 generation progeny should be homozygous for this mutant and, therefore, recognizable. Where these recessive mutants are sex-linked, the procedure is somewhat modified. Students should work out for themselves a suitable breeding programme. Naturally, if you are going to undertake this sort of project, you must be very familiar with the normal phenotype of the fly you are studying. Several hours spent in observing and drawing a normal specimen will prove useful when it comes to examining F_2 generation progenies.

The action of natural selection upon a recessive gene can be shown in populations of *D. melanogaster* maintained in population cages. Details of how to set up such an experiment using the recessive sex-linked mutant white-eyed (w) are given in the Investigations. Figure 24 gives the data collected by Reed and Reed (1949). The solid line shows the gradual elimination of white-eyed individuals from the population by natural selection. The crucial stage in the life of the fly, at which selection appears to act against white-eyed individuals, was discovered to be that of courtship. Both normal and white-eyed females discriminate against white-eyed males when given a choice. White-eyed males were only 75% as successful in mating compared with their normal counterparts. Using these data, Reed and Reed calculated a curve (fig. 24, dashed line) for the expected decrease in percentage of white-eyed flies from the initial 50% until extinction. The observed and expected curves were found to agree remarkably well suggesting that, in this case, the selective mating on the part of the female was the main factor in the elimination of the mutant from the population.

5 CHROMOSOMES

In the salivary glands of insects belonging to the order Diptera (flies, mosquitoes, and midges) the chromosomes have become greatly enlarged by a process of replication. These replicates are lined up alongside one another and do not separate. This process of chromosome multiplication, called *polyteny*, produces chromosomes that are many times thicker than usual and magnifies any differences in density along their lengths. Such density differences mean that polytene chromosomes display a precise pattern of banding which enables us to identify any one chromosome or any specific part of a chromosome. In most animals the homologous pairs of chromosomes usually remain separate except during *meiosis* (cell-division leading to gamete-formation). However, in the salivary-gland chromosomes, the two homologous chromosomes of a pair never separate. This situation is known as *somatic pairing*.

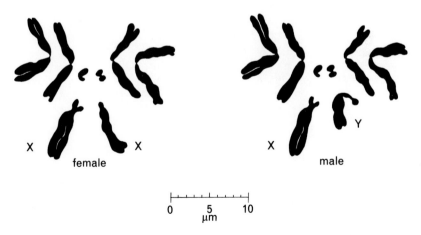

Fig. 25 Chromosomes from dividing cells of larval brain of *Drosophila melanogaster*.

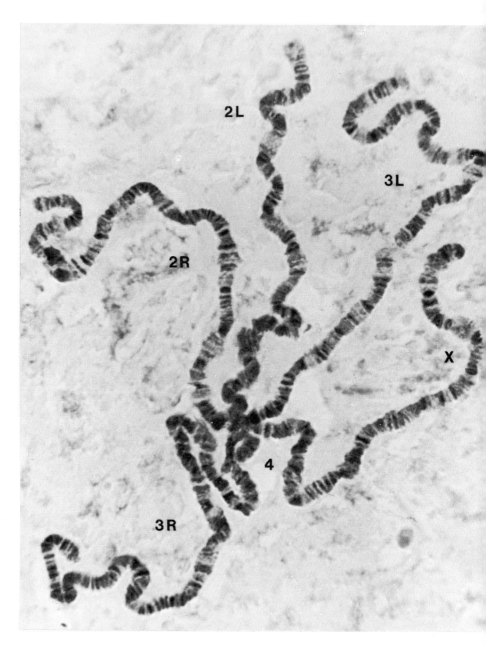

Fig. 26 Salivary-gland chromosomes of *Drosophila melanogaster* female larva.

In *D. melanogaster* there are four pairs of chromosomes. Figure 25 shows the chromosomes from dividing cells of a larval ganglion. Those of a female are on the left and those of a male are on the right. Both have a pair of very small dot-like chromosomes (chromosome 4) and two pairs of large V-shaped chromosomes (chromosomes 2 and 3). These three pairs are known as the *autosomes*. In addition, both have a pair of sex-chromosomes. In the female these are similar in size and shape (X-chromosome or chromosome 1); in the male there is one X-chromosome and one J-shaped Y-chromosome. These chromosomes from the brain of the larva are extremely small (note the 10 μm scale in the figure). However, the salivary-gland chromosomes obtained from third instar larvae are approximately 100 times the length of these chromosomes (fig. 26). In the figure, the banding patterns can be seen clearly and the four chromosomal units (each unit being an homologous pair of chromosomes) are labelled. 2L and 2R are the left and right arms of the second chromosome. Similarly, 3L and 3R are the left and right arms of chromosome 3. All the chromosomal units radiate out as if from a common centre. This region, known as the *chromocentre*, is formed by the mutual attraction and close association of that part of each chromosome known as the *centromere*. It is positioned at one end of the X-chromosome (*telocentric*) but towards the centre of chromosomes 2 and 3 (*metacentric*).

A feature of great interest to a number of *Drosophila* workers has been the presence of *chromosomal inversions*. You will be aware already from the previous chapter that chromosomes frequently break and rejoin. The process of recombination already described depends upon this property. If a chromosome breaks in two places simultaneously and the middle section then revolves through 180° before rejoining with the broken portion on each side, the chromosome will have an inverted section in its length. In Figure 27A the middle section becomes inverted, giving rise to a chromosome with an inverted sequence of genetic material (fig. 27B). If a fly is homozygous for a particular sequence, then the close pairing of homologous chromosomes that occurs during meiosis (and permanently in the salivary glands) can proceed without any difficulty. However, if a fly is heterozygous for a particular sequence, then homologous partners have different arrangements of genes and pairing difficulties arise. Close pairing can be achieved only by forming a loop as in Figure 27C, to bring about maximum pairing of homologous points along the two chromosomes. The presence of such looped structures in the salivary-gland chromosomes can quite easily be identified and the exact boundaries of the inverted segment recorded (fig. 28).

In a number of *Drosophila* species many inversions have been recorded and named. Sometimes, as in *D. pseudoobscura*, a species from

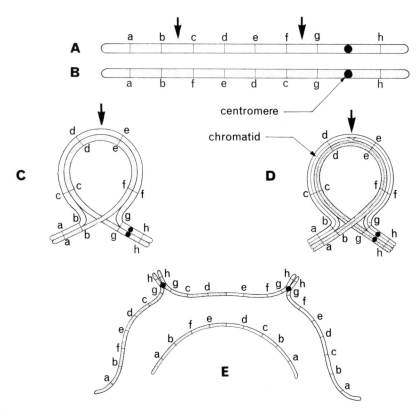

Fig. 27 Chromosomal inversion; **A** ancestral chromosome with the original sequence of genetic material; **B** chromosome with a new sequence of genetic material caused by the inversion of genes c-f (arrowed in **A**); **C** pairing of ancestral and modified chromosome; **D** the same two chromosomes early in meiosis, when a cross-over (arrowed) has occurred between two of the chromatids; **E** the products of meiosis.

the United States, only one of the chromosomes is present in several different forms. On the other hand, in the European *D. subobscura*, all five of the large chromosomes carry inversions. In many cases several different inversions can be present in a population of flies. Such a species is said to exhibit *chromosomal polymorphism*.

Having mentioned inversions, it seems necessary to ask what value are they to the species, particularly in populations which exhibit poly-

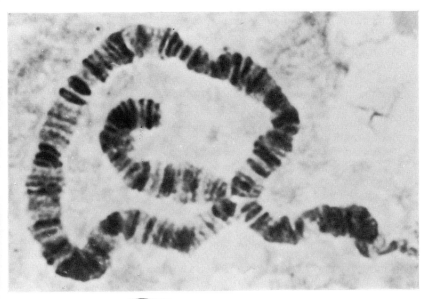

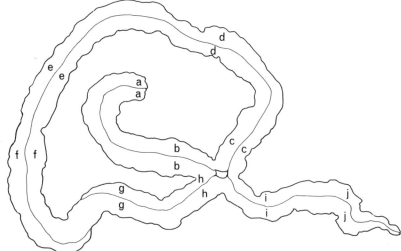

Fig. 28 Photograph of a salivary-gland chromosome of *Drosophila melanogaster* showing an inversion loop. The interpretative diagram shows the two homologous chromosomes, one of which has the ancestral sequence (abcdefghij) while the other has an inverted sequence (abhgfedcij).

morphism? This question is best answered by considering what effect inversions have upon recombination. Take the paired homologous chromosomes in Figure 27. Imagine that a cross-over occurs at the point indicated by the arrow in Figure 27C; that is, within the inverted segment. In the early part of meiosis, at the four-chromatid stage, the cross-over would produce the configuration shown in Figure 27D. The four chromosomes produced are illustrated in Figure 27E. The two parental types are normal but the recombinant types suffer from a duplication of genetic material on the one hand and a deficiency on the other. As a result, neither of the recombinant chromosomes behaves normally in cell-divisions and the only viable products of meiosis are those containing the parental combination of genes. Inversions, then, effectively prevent the recombination of genes.

Since genes do not act independently but in a co-ordinated manner to produce the final phenotype, it is not surprising that certain combinations of genes will interact together more favourably than others. Natural selection will tend to preserve such favourable combinations; recombination will tend to destroy them. If such a co-ordinated (*co-adapted*) group of genes are included within an inverted segment of chromosome, no recombination can occur between them. So, inversions preserve co-adapted gene-complexes.

Different complexes will be favoured under different environmental conditions. One might expect, therefore, that the frequencies of the various inversions found in a *Drosophila* species will vary in different parts of the species' geographic range.

Much work has been done on the inversion types of *D. subobscura*. In this species there are six pairs of chromosomes: five long chromosomes and one small dot-like chromosome. All five of the long chromosomes have forms containing inverted sections and the total number of inversions that have been found is now more than 40. To illustrate the geographic variation of chromosomal inversions, we will use one of these chromosomes as an example, although similar variations are shown by all of them. We will choose the sex-chromosome, usually called chromosome A in this species. One particular sequence shown by this chromosome has been named for convenience the standard form (A_{st}); two alternative forms $(A_1$ and $A_2)$ showing inverted sequences have also been found.

Prevosti (1966) compared five populations of *D. subobscura* from Edinburgh (Scotland), Lagrasse (France), Barcelona (northern Spain), Ribarroja (central Spain), and Malaga (southern Spain). The frequencies of the three possible forms of chromosome A are shown in Figure 29 (*top*). Certainly, with forms A_{st} and A_2, a change in frequency from northern Europe to southern Europe can be seen clearly. In Figure 29 (*bottom*) a

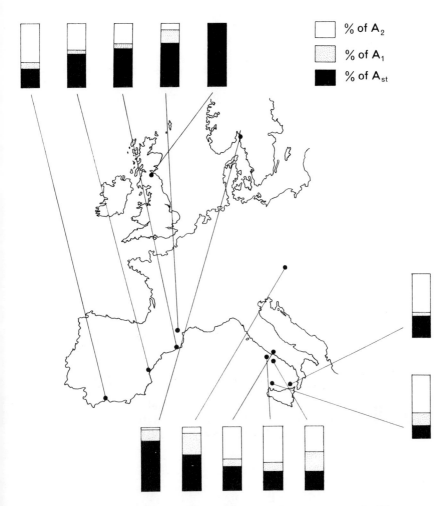

Fig. 29 Frequency of inversion types in chromosome A (the sex chromosome) of *Drosophila subobscura*. A_{st} is known as the standard form, and A_1 and A_2 are different inversions of it.

similar change in the frequency of these chromosomal types is seen, although these populations came from a little further east in Europe. While the standard form shows a decline in frequency from north to south, this is accompanied by a corresponding increase in the frequency of A_2. We conclude that flies possessing the standard arrangement of chromosome A are favoured in northern Europe, while a Mediterranean environment is more favourable to those possessing the sequence A_2.

Another chromosomally polymorphic species is *D. funebris*, and some interesting facts concerning the distribution of these inversions have been collected by Dubinin and Tiniakov in Russia. Their findings are not set out in the form of frequencies of inversion types but rather as the percentage of individuals in a population that are heterozygous for one or more inversions. This species has six pairs of chromosomes and five inversions have been found. The inversion heterozygotes appear to be many times more frequent in the populations of cities than in more rural areas. The best data were collected in and around Moscow and are set out in the following table:

Populations of *D. funebris*	Inversion heterozygotes
in central part of Moscow	88·1%
in residential part of Moscow	55·5%
near Moscow suburbs	12·1%
in villages 20-200 km from Moscow	1·8%
in villages 200-500 km north of Moscow	0·0%

Other Russian cities were found to contain polymorphic populations of *D. funebris* and, as might have been predicted, it was found that larger cities and towns contained populations which were more polymorphic.

In a book of this size little more can be said about inversions; for further reading, there is an excellent account of similar work done on North American species by Dobzhansky (1951, 1970).

6 BEHAVIOUR

All insects show an interesting array of behavioural responses to stimuli of various kinds; *Drosophila* is no exception. All the ones described in this chapter can be examined easily in the laboratory.

Courtship

Before copulation, *Drosophila* perform a sequence of activities usually referred to as courtship. An individual male can mate many times; females store the sperm from this single mating in the seminal receptacles and spermathecae. As each egg passes down the uterus, a small amount of this stored sperm is released to bring about fertilization. Even if the store of male gametes is eventually exhausted, the female may continue to lay unfertilized eggs. Thus, if you wish to observe courtship behaviour, you must obtain virgin females. These should be about four days old because young immature virgin flies will refuse to mate also.

In all species the courtship itself is preceded by the male tapping the abdomen of the female with his foreleg. This appears to be a means of identifying a female of his own species. If the female belongs to another species the male does not normally continue with the courtship. Such mechanisms of species recognition are essential if species are to remain distinct and separate. They are usually referred to as *isolating mechanisms*, and prevent the formation of species hybrids.

In *D. melanogaster*, after the tapping procedure has shown that the female is of the correct species, the male follows her and extends one wing to an angle of 90° to his body (fig. 30). This wing, which is normally that nearest to the female's head, is vibrated periodically for several seconds. If the female is receptive, she stops running and allows the male to lick her genitalia with his proboscis, mount her, and then copulate.

Males can make mistakes in the tapping procedure, which is carried out only once at the very beginning of courtship. Also, if the sequence of courtship is interrupted for some reason, the male may lose sight of the first female and transfer his attentions to another. In a mixed species population it sometimes happens that this second female is of another

Fig. 30 The courtship of *Drosophila melanogaster* showing the wing
vibrations of the male.

species. Even though the male continues to court such a female she will
not necessarily allow copulation. Obviously, she is not satisfied with the
male in some way and recognizes that he belongs to another species.

It was originally suggested that the female recognized the male
visually, but since mixed populations of *Drosophila* kept in the dark mate
quite successfully and without hybridization, this explanation seems
unlikely. Quite recently, Bennet-Clark and Ewing (1968) working at the
University of Edinburgh, have discovered that, during courtship, male
Drosophila emit a noise, so recognition by the female may well be auditory.
Apparently, the sound or song is produced by the upstroke of the male's

wing during the vibration stage of the courtship sequence. The song normally consists of a train of sound pulses, with each pulse followed by a period of silence. In *D. melanogaster* each pulse lasts for about 0·003 s and is repeated 30 times per s.

In *D. simulans*, a species closely related to *D. melanogaster* and almost identical in external appearance, the courtship pattern also appears to be superficially the same. It differs only in that *D. simulans* is a less active species and occasionally exhibits an additional wing display called *scissoring*. This consists of opening and closing the wings several times. While performing these movements the male normally stands in front of the female so that it is possibly a visual display. Males of *D. simulans* also produce a song with the same sound pulse as *D. melanogaster*. However, it is repeated only 20 times per s. Reproduced in Figure 31 are oscilloscope patterns for both species showing these differences which, according to Bennet-Clark and Ewing, may be "codes" which allow females to recognize males of their own species.

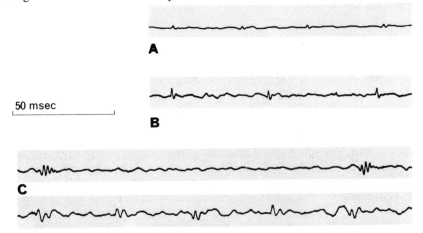

Fig. 31 Oscilloscope records of courtship "songs" of *Drosophila*; **A** *D. melanogaster*; **B** *D. simulans*; **C** *D. ambigua*, upper record showing flicking movements, lower record showing vibrations.

The importance of the song produced by the males has been shown experimentally by these workers. Males that have had their wings removed have been shown to be very unsuccessful in obtaining mates, even though their courtship remains persistent and they attempt to copulate just as frequently as do normal males. In another experiment, female flies were introduced into a mating chamber along with wingless males. The

73

simulated song of the male was then played. The result was that the number of successful copulations increased, although not to the level achieved by normal males. Since the normal wing vibrations cause a movement of air, a further series of experiments were carried out in which not only was the song played to the flies but also a current of air was passed through the chamber. The result was spectacular; the number of successful copulations increased to that achieved by normal males.

All the species examined so far appear to have their own distinctive song, with the exception of *D. subobscura*. This species is interesting because it does not mate in the dark. The typical sequence of events in *D. subobscura* upon placing a virgin female and male together in an observation chamber is as follows. Once the male sees the female, he approaches her performing a rapid series of flicks with his wings. After the initial tapping with the forelegs he circles round so as to face her head to head, with his proboscis extended towards her. At this point, both sexes perform a side-stepping dance facing one another, during which the male raises and spreads his wings but does not vibrate them. The female now stands still and the male moves round to her posterior end and copulates. During the whole of the display, Ewing and Bennet-Clark did not detect any sound production of any kind and it appears, therefore, that in this species the important courtship stimuli are visual rather than auditory.

An example of a courtship routine that is a little more complex is that of *D. ambigua*. The beginning and end of the sequence of courtship behaviour are similar to those of *D. melanogaster*. It is the wing display that is more complex. Both wings are flicked out from the body in such a way that each flick extends them a little further until an angle of 85° to the body is reached. During this movement the activity of one of the wings becomes progressively less and less until it finally stays in the normal resting position. The movements eventually change into vibrations similar to the ones described for *D. melanogaster*. The recorded song of *D. ambigua* has two quite different components: one corresponding to the initial wing movements and one to the vibrations. Both types of sound are shown in Figure 31C.

Throughout the whole of this description of courtship behaviour nothing has been said about how the female receives the sound stimulus from the male. Aubrey Manning, also working at the University of Edinburgh, has shown that if the aristae attached to the flies' antennae are removed or glued down, the female fly becomes unreceptive. Observations of females subjected to loud pure sounds show that the aristae and parts of the antennae vibrate with a natural resonance of about 200 cps. The aristae may then be the female's sound receivers.

So far as producing sound herself the females of all species studied so far produce one very similar sound. This consists of a long and very loud buzz, and is used mostly by very young and unreceptive virgins to repel unwelcome males. Since all females appear to make the same kind of sound, it will be effective not only against males of their own species but also those of other species. Such a universal "no" signal has quite obvious advantages.

Orientation to light and gravity
In general it can be said that *Drosophila* move towards light (*positive phototaxis*) and away from gravity (*negative geotaxis*). This statement requires some qualification, however.

The reactions of *Drosophila* to gravity and to light are under genetic control and can be modified, therefore, by selection. Also, the form of the response, particularly towards light, can vary depending on the species and the environmental situation. A number of experiments performed by Lewontin (1959) on the North American species *D. pseudoobscura* illustrate this point. A length of glass tubing, shaped like a letter Y, was laid horizontally with light coming in through a window on one side only. A square of cardboard placed between the arms of the fork kept one arm in light and the other in shadow. A single fly was placed in the apparatus so that it could walk down the main tube and enter either of the two side-arms. Its choice was scored when it had walked halfway down one arm. Each fly was tested twice, the Y-tube being turned round between the two runs so that the arm that had been in shadow was now in the light. Thus, any inherent difference in the tube or possible scent-tracks produced by the flies would not bias the results. Under these conditions, Lewontin found that *D. pseudoobscura* avoided the light.

Lewontin repeated the experiment, but this time he placed 10 flies in the starting tube at one time. Since it was impossible to recognize each individual fly the result was scored in a different way. Each test lasted for 5 minutes, after which time the numbers of flies in the two arms of the tube were counted. Again, there was strong evidence of negative phototaxis, but Lewontin noticed that many flies went to the light side at first and then moved back into the shaded side. He thought, therefore, that the level of excitement of the flies might be important.

In order to test this hypothesis he conducted another group of experiments using similar techniques but a different testing chamber. He substituted a large glass jar for the Y-tube; here, the flies could fly as well as walk. The body of the jar was divided internally by a piece of cardboard and the jar laid on its side. The vertical cardboard divider left half the glass jar illuminated and kept the other half in shade. A number of holes in the cardboard partition allowed flies to move freely from the light to the shade and *vice versa*. When flies were introduced into the neck of the jar and left undisturbed the same result was obtained: that is, the flies showed negative phototaxis. However, in a final series of runs, 50 flies were released into the apparatus and the jar was then banged on the table. In a high level of excitement the flies flew into the light side, thus displaying the strong photopositive behaviour familiar to all those who work in laboratories with *Drosophila*. As their level of excitement decreased, the flies walked back into the shade.

The strong positive phototaxis at high levels of activity would appear to be an escape reaction. In the case of *D. pseudoobscura* the reversal of this at low levels of excitement may be an adaptation to the dry environment that this species frequents. Dark places, such as the undersides of fallen leaves and the interstices of decaying logs, are usually more moist than open places and are, therefore, sought out by the flies. In this connection it is interesting to note that *D. persimilis*, a closely related species which lives in moister, cooler climates is photopositive even at low activity levels.

Food preference
In the laboratory, fruit flies are attracted to a variety of organic compounds that are found naturally in fermenting fruit. These include amyl and ethyl alcohol, acetic and lactic acid, and ethyl acetate. In many cases, mixtures of some of these compounds are more attractive to the flies than the pure substances.

Attraction to various food substances can be demonstrated under more natural conditions, too. Okada (1962), working in Japan, investigated the degree to which flies of different species preferred sap to fruit and how this food preference changed with environmental conditions. A mass of decaying fruit was spread on the ground about 10 m from the base of a bleeding oak. Thus, visiting *Drosophila* had to choose between sap and fruit. Although some species went to both, many fed mainly from one or the other. With the increase of the fly population in the warmer season, however, more of the flies used both kinds of food. In other words, with increased density flies become less particular about the food they eat. Another interesting observation was that the sap attracted flies that were mainly black or dark brown in colour. Those caught at the fruit were mostly yellowish or pale brown species. It seems possible that the dark coloration of the sap-feeders may help to conceal them against potential predators because it matches the dark bark of the tree trunks.

Part of this food preference might be the result of conditioning. For example, only 35% of flies reared on standard laboratory media are attracted by the odour of peppermint. However when flies are reared on a medium containing 0·5% of oil of peppermint, 67% of the emerged adults find the odour attractive.

Of course, in the wild it is not the sap or fruit itself that *Drosophila* feed upon but the micro-organisms, particularly yeasts, that grow upon these carbohydrate substrates. Experiments have shown that fruit flies have a preference for certain kinds of yeasts. In the laboratory, both larvae and adults were offered a choice of food and the particular preference, as indicated by positive movement towards the food, was recorded (see Investigations for experimental details). Both larvae and adults show preferences under these conditions and these differ from species to species. Furthermore, the preferences are not necessarily the same in larvae and adults for any one species. In many cases, the yeasts preferred by the larvae were those that also best supported growth. This means that should the female lay her eggs in an unfavourable site, the larvae have a limited ability to search out those areas in which they will develop best.

Experiments by Dobzhansky and his colleagues (1956) in the Yosemite region of California have shown that this differential attractiveness to various yeasts occurs in the wild also. Bait, placed in cans, was set out at fixed locations. In most cases the distance between neighbouring cans was 10 to 20 m so that they "competed" with each other for the flies they attracted. The bait in each can was inoculated with one of 15 species of yeasts, all but two of which had previously been isolated from crops of flies caught in the region. Again, there was a difference between the yeast preference of different species. Undoubtedly this will lessen the competition for food that must go on between the species, and so enable a number of *Drosophila* species to live together in the same area by exploiting a slightly different food-source.

7 IDENTIFICATION

Adult Diptera (true flies) are easily recognized by their single pair of wings. By this criterion they can be distinguished from nearly all other winged insects. The hind pair of wings are replaced by a pair of drumstick-like structures called halteres.

In both the number of individuals and the number of species the Diptera group is a large one. In Britain alone more than 5000 species have been recorded. Within the family Drosophilidae, however, only 52 species are found in the British Isles and, of these, 32 belong to the genus *Drosophila*. This number includes 2 "casual species" which have so far not been recorded from the mainland of Britain but are frequently encountered on ships in British ports. The key that follows is concerned only with the most common 22 of the remaining 30 species. A complete key to the British Drosophilidae has been published by Fonseca (1965).

The characteristics of the family Drosophilidae are briefly as follows: they are small flies, attracted to fermenting substances; they have a very characteristic slow, ponderous, hovering flight. In colour they range from yellow, through brown, to black; typically, they have bright red eyes; they have no hairs or bristles on the mesopleuron. In the wing the costal vein is broken twice, once near the humeral cross-vein and again close to the top of vein I. Vein I is short and the anal cell and vein VI are always present. There is no trace of a cleft on the outer (lateral) surface of the second segment of the antenna. The third segment of the antenna is rounded and short. The arista is generally *plumose* (much branched), with a fork at the end called the *apical fork*, formed from the main stem and the last branch. On the head, the postvertical bristles are usually *convergent* (pointing towards each other) but the 2 or 3 orbital bristles always curve down or up, never inwards and the middle (shortest) orbital bristle is always set somewhat closer to the eye than are the others.

Within the family Drosophilidae the genus *Drosophila* can be distinguished by the following additional characters: the eyes are covered with a dense short pile of hairs, the postvertical bristles are well developed and usually cross over each other, and there are three orbital bristles. On the thorax there are two pairs of dorsocentral bristles. The majority of *Drosophila* species have six or eight rows of acrostichal bristles which lie just in front of the anterior pair of dorsocentral bristles. The three British *Drosophila* species possessing only four rows of acrostichal bristles can be distinguished from the closely related genus *Scaptomyza* (which also have four rows of bristles) by the mid-sternopleural bristle which, in *Scaptomyza*, is not the shortest of the three.

Within the genus *Drosophila* at least eleven smaller divisions or sub-genera are now recognized and within certain of these the individual species are arranged in groups. A full list of the sub-genera and species groups is not given here; for a useful description see Patterson and Stone (1952).

Key to common British species of *Drosophila* (after Frydenberg, 1956)
Note: The body colorations described in the following key vary depending upon the method of viewing. This can be carried out in air or in alcohol, with the light source above or to the side of the specimen. All the colours described here (including the coloured plates) refer to freshly killed mature specimens placed in 70% alcohol and illuminated from above. The darker species particularly change their colour quite markedly when transferred from air to alcohol. In these cases, the colour observed in air is enclosed in square brackets.

1 . Four rows of acrostichal bristles (fig. 32A);
   three sternopleural bristles, the middle one
   intermediate in length   2
   Six or 8 rows of acrostichal bristles (fig. 32B, C);
   2 or 3 sternopleural bristles; if 3, middle one always
   much shorter than anterior bristle   3

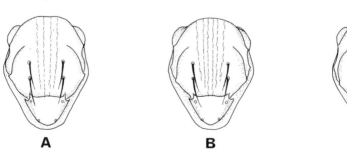

      A   **B**   **C**

Fig. 32 Dorsal view of thorax showing acrostichal bristles; **A** 4 rows; **B** 6 rows; **C** 8 rows.

2   Male: claspers very big, broadest ventrally;
    female: ovipositor plate with a row of small
    teeth all the same size   *D. fenestrarum* p. 94
    Male: claspers smaller but prominent, pointed
    ventrally with triangular projection at base;
    female: ovipositor plate with 5 to 7 teeth, 2
    longer and stouter than the rest   *D. andalusiaca* p. 93

3   Mesonotum brown [grey with a greenish tinge];
    almost all bristles and hairs on mesonotum
    inserted on a dark brown [black] spot (plate 7)   *D. hydei* p. 95
    Mesonotum yellow, brown, or black; no special
    colour around bases of mesonotal bristles and hairs   4

4   Mesonotum yellow or yellow-brown,   5
    Mesonotum darker brown or black   14

5   First femur on the inner side with a row of about
    10 small black spines; veins clouded at wing apex
    and on posterior cross-vein; abdomen yellow
    with 2 big, dull, dark brown [black] triangles on
    each tergite (fig. 33; plate 5)   *D. immigrans* p. 96
    None of the above characteristics   6

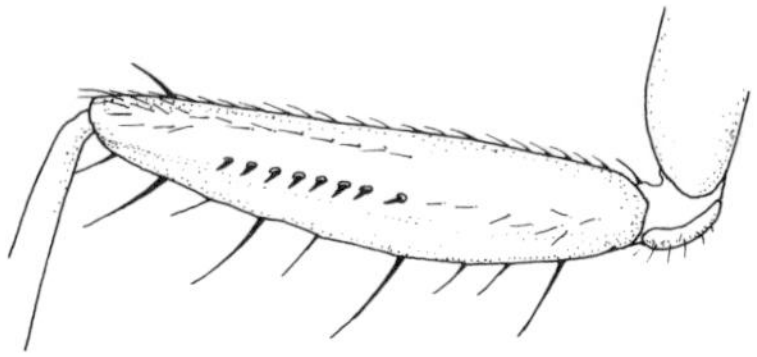

Fig. 33 *Drosophila immigrans*;
first femur showing row
of small black spines.

6   Mesonotum yellow with 5 dark brown [black]
    longitudinal stripes, middle stripe bifurcating
    posteriorly; yellow pleura with 2–3 dark stripes
    (plate 2)   *D. busckii* p. 93
    Mesonotum without 5 distinct stripes, often
    without any pattern at all   7

7   Yellow abdominal tergites with dark brown
    [black] band along posterior margins; bands
    never interrupted in median line (they are most
    often broadest here); 5th and 6th tergites in the

male all dark brown [black]; males with sex combs   8
Posterior bands on at least the 4 most anterior
tergites interrupted in the median line   9

8   Males: genital arch with small hook-shaped process;
both sexes: width of cheeks (measured
from lowest point of eye to margin of cheek)
about 1/6th of the greatest diameter of the eye;·
maxillary palps most often with 3 stouter bristles
on their outer end (fig. 34A, C, E; plate 3)   *D. melanogaster* p. 97
Males: genital arch with large, hood-shaped
process; both sexes: the width of the cheek is
less than 1/6th of the greatest diameter of the
eye; maxillary palps most often with 2 stouter
bristles on their outer end (fig. 34B, D, F)   *D. simulans* p. 98

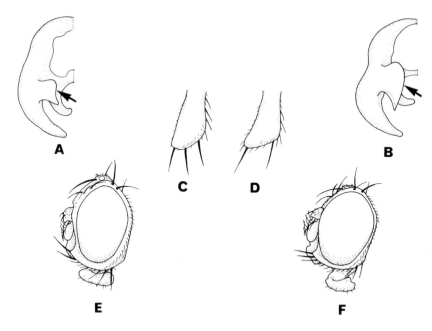

Fig. 34 *Drosophila melanogaster*; **A** left half of male genital arch;
C maxillary palp; **E** lateral view of head.
Drosophila simulans; **B** left half of male genital arch; **D**
maxillary palp; **F** lateral view of head.

9 Wings completely unclouded   10
  Wings clouded or, at least, much more darkly
  pigmented on the 2 cross-veins (fig. 35)   11

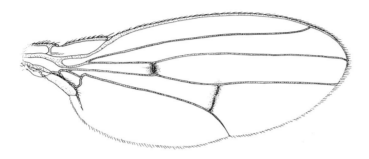

Fig. 35 Wing of *Drosophila transversa* showing pigmented cross-veins.

10 Eight rows of acrostichal bristles between
  dorsocentral rows   *D. confusa* p. 94
  Six rows of acrostichal bristles between
  dorsocentral rows   *D. cameraria* p. 94

11 Dark brown [black] bands on posterior margins
  of abdominal tergites interrupted both in the
  median line and laterally giving 4 dark brown
  [black] spots on each tergite (fig. 36A)   *D. transversa* p. 99
  Posterior bands on tergites usually interrupted
  in median line only, giving 2 dark brown [black]
  spots on each tergite (fig. 36B, C, D)   12

12 Second, 3rd, and 4th abdominal tergites each
  with a pair of large, more or less triangular dark
  brown [black] spots with base on hind margin
  and apex reaching fore-margin (fig. 36D); cross-
  vein darkly pigmented rather than clouded   *D. histrio* p. 95
  Dark brown [black] markings on tergites 2, 3,
  and 4 more rectangular in shape and occupying
  less than half the length of tergite; cross-veins
  clouded (fig. 36B, C)   13

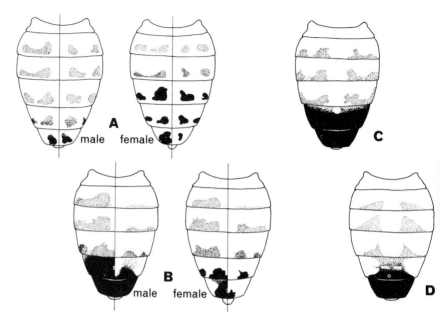

Fig. 36 Abdominal colour patterns; **A** *Drosophila transversa*, male and female; **B** *Drosophila phalerata*, male and female; **C** *Drosophila kuntzei*, male; **D** *Drosophila histrio*, male. In **A** and **B**, the left hand side of each drawing is of a specimen showing a strongly developed pattern and the right hand side is of a specimen with a weakly developed pattern.

13 Male: fringe of much longer hairs on anterior face of apical third of front metatarsus and along entire length of second tarsal segment; claspers square with peg-like teeth of clasper comb becoming more bristle-like at lower end; female: ovipositor guide, in profile, broadly rounded at tip with upper and lower margins almost parallel (fig. 37A, B, D; plate 4)   *D. phalerata* p. 97

Male: no fringe on front tarsi; claspers triangular and all teeth of clasper comb peg-like; female: ovipositor guide more triangular and more narrowly rounded at tip, with upper and lower margins making an angle of about 30° (fig. 37C, E)   *D. kuntzei* p. 96

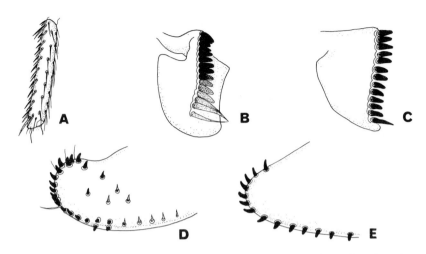

Fig. 37 *Drosophila phalerata*; **A** left metatarsus of male showing fringe of hairs; **B** male's left clasper; **D** female's right ovipositor guide.
Drosophila kuntzei; **C** male's left clasper; **E** female's right ovipositor guide.

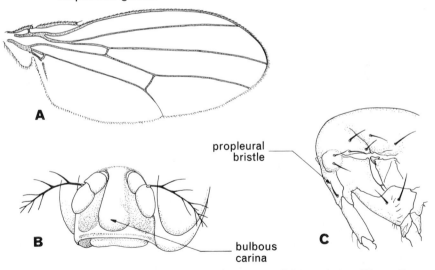

Fig. 38 *Drosophila deflexa*; **A** wing; **B** carina; **C** lateral view (from the left side) of anterior half of thorax, showing position of propleural bristle.

14 Lower part of carina bulbously swollen; pair of
   inconspicuous prescutellar bristles present; small
   distinct bristle on propleuron; tip of abdomen in
   living specimens withdrawn and genitalia often
   invisible; wings yellow, short (fig. 38)   *D. deflexa* p. 94
   Carina nose-shaped, never bulbous; no
   prescutellar bristles and no pleural bristle;
   external genitalia usually well exposed and
   clearly visible in living specimens   15

15 Six rows of acrostichal bristles   16
   Eight rows of acrostichal bristles   17

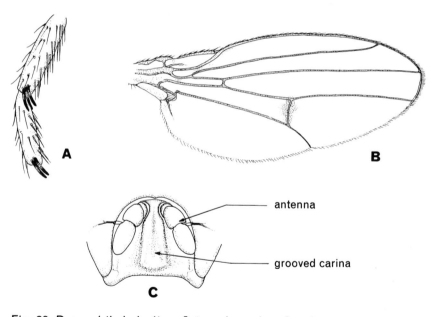

Fig. 39 *Drosophila helvetica*; **A** tarsal combs of male.
Drosophila littoralis; **B** wing; **C** anterior view of head showing
carina.

16 Posterior cross-vein intensely clouded;
   mesonotum dark brown [greyish black];
   abdomen dark brown; carina with a longitudinal
   groove; large species (2 to 4 mm); males without

tarsal sex combs (fig. 39B, C; plate 1)   *D. littoralis* p. 96

Posterior cross-vein unclouded; carina not
grooved; very small species (1·25 to 2·00 mm);
males with 2 tarsal sex combs, upper comb 2 to 5
teeth, lower comb 1 to 3 teeth (fig. 39A)   *D. helvetica* p. 95

17 Mesonotum red-brown; 3 sternopleural bristles;
    abdomen dark brown [black] but with yellow
    band along anterior margin of at least first 4
    tergites; yellow band broadest in the median
    line; males without sex combs on forelegs; large
    species (3 to 4 mm; plate 6)   *D. funebris* p. 95

    Mesonotum brown, black, or black with brown
    stripes; 2 sternopleural bristles; abdominal
    tergites dark brown [black] without anterior
    bands; males with 2 sex combs on each foreleg;
    small species (2 to 3 mm; plate 8)   18

18 Males   19
    Females   23

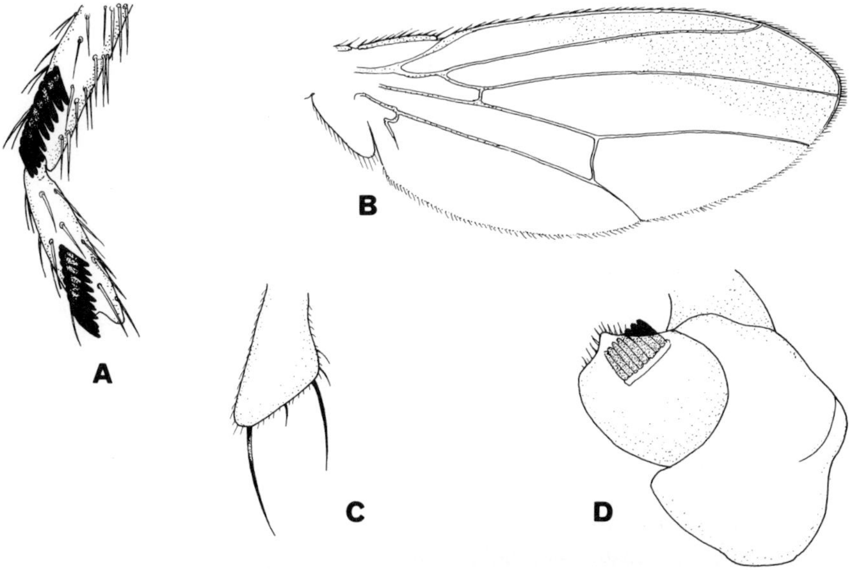

Fig. 40 *Drosophila tristis*; **A** tarsal combs of male; **B** wing of male;
    C maxillary palp; **D** right genital arch and clasper of male.

19 Wings strongly clouded at the tip; palps with
   2 equally long bristles; upper tarsal comb with
   9 to 12 teeth, lower with 8 to 11 teeth; external
   process of genital arch obtusely wedge-like and
   short; claspers very round and cup-shaped,
   containing the clasper comb of 6 to 8 teeth (fig. 40)   *D. tristis* p. 99
   Wings unclouded or faintly clouded around
   end of vein II; palps with 1 strong bristle; if
   2 bristles, then of unequal length   20

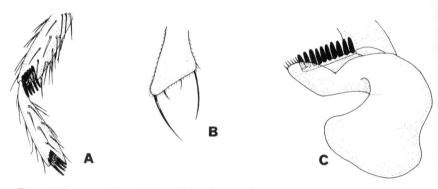

Fig. 41 *Drosophila subsilvestris*; **A** tarsal combs of male; **B** maxillary
   palp; **C** right genital arch and clasper of male.

20 Palps with a strong terminal bristle and a weak
   subterminal one; upper tarsal comb with 4 to 6
   teeth, lower comb with 3 to 5 teeth; mesonotum
   dark brown without any pattern; external process
   of genital arch shorter than clasper; latter very
   long and irregular, containing clasper comb of
   8 to 10 teeth (fig. 41)   *D. subsilvestris* p. 98
   Both tarsal combs with at least 6 teeth;
   with or without colour pattern on mesonotum   21

21 Palps with 1 strong terminal bristle, with often
   a considerably smaller subterminal bristle; upper
   tarsal comb with 6 to 10 teeth, lower with 6 to 8
   teeth; mesonotum with longitudinal stripes;
   external process of genital arch long and wedge-
   shaped; claspers long and graceful containing the
   clasper comb of 7 to 10 teeth (fig. 42; plate 8)   *D. obscura* p. 97

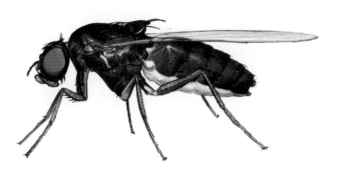

Plate 1   *Drosophila littoralis* (female) from Ravaniemi, northern Finland; length 3–4 mm.

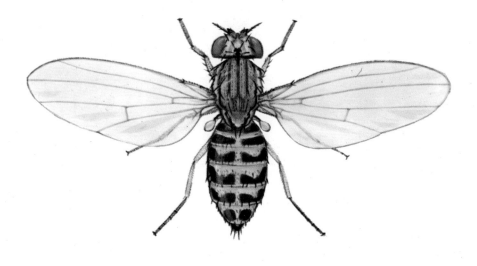

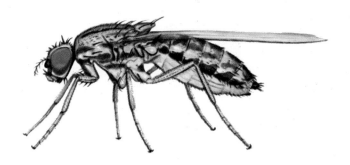

Plate 2   *Drosophila busckii* (female) from Newcastle-upon-Tyne; length 2–2·5 mm.

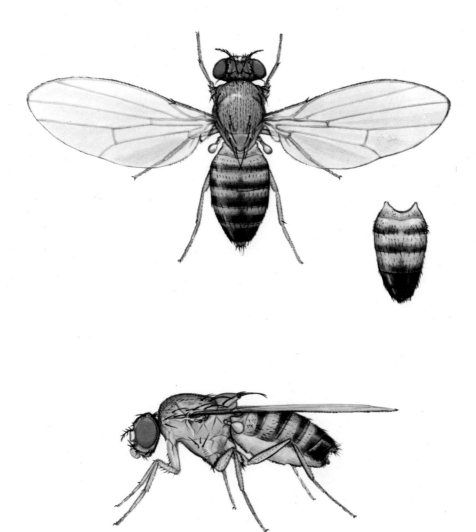

Plate 3   *Drosophila melanogaster* (female, and male abdomen) from Rogate, Sussex;
length   2–3 mm.

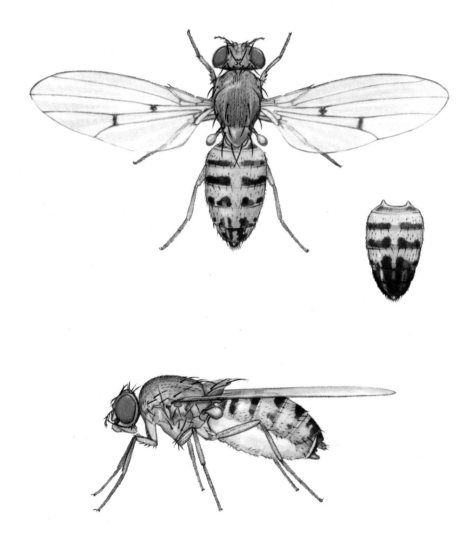

Plate 4 *Drosophila phalerata* (female, and male abdomen) from Woodchester Park, Gloucestershire; length 3–4 mm.

Plate 5 *Drosophila immigrans* (female, and male abdomen) from Newcastle-upon-Tyne; length 3–3·5 mm.

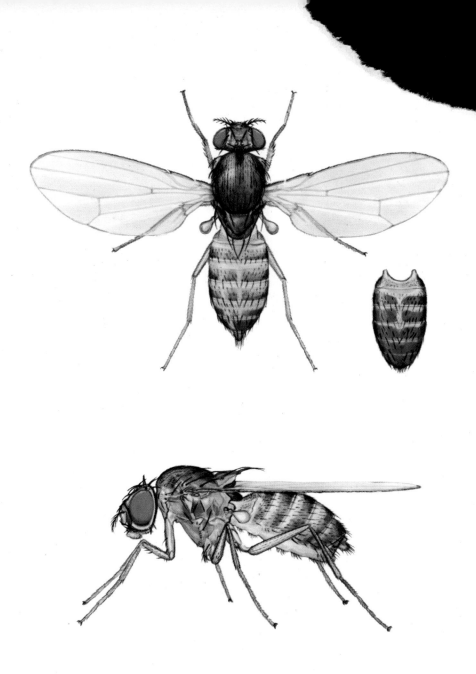

Plate 6 *Drosophila funebris* (female, and male abdomen) from Manchester; length
3–4 mm.

Plate 7 *Drosophila hydei* (female) from Kuopio, central Finland; length 3–4 mm.

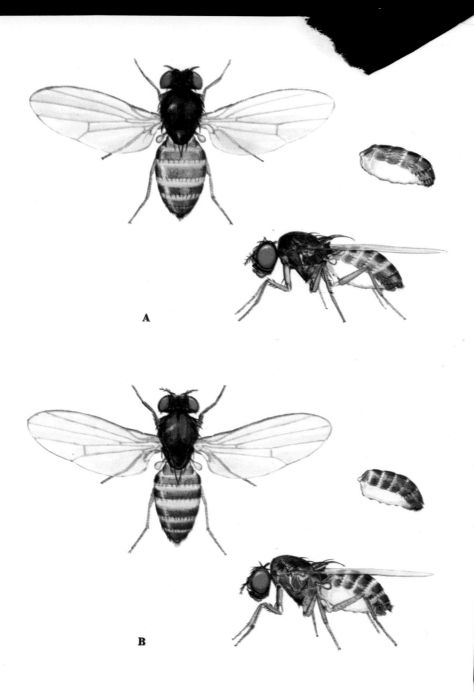

Plate 8 **A** *Drosophila subobscura* (female, and male abdomen) from Rogate, Sussex; length 2–3 mm; **B** *Drosophila obscura* (female, and male abdomen) from Kolmperä, Espoo, southern Finland; length 2–3 mm.

Tarsal combs very long; upper comb with at
least 8 teeth; mesonotum without any pattern
(plate 8)   22

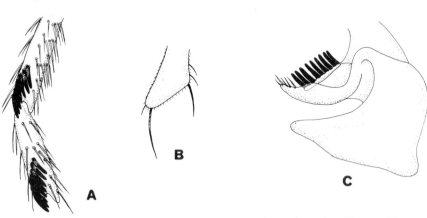

B

C

A

Fig. 42 *Drosophila obscura*; **A** tarsal combs of male; **B** maxillary
palp; **C** right genital arch and clasper of male.

22 Palps with only 1 conspicuous bristle;
    upper tarsal comb with 10 to 15 teeth, lower
    with 9 to 13 teeth; external process of genital
    arch rounded and bulging at base and drawn out
    into a thin projection; clasper large and cup-
    shaped, laterally compressed and containing a
    very short and square-looking clasper comb with
    6 to 8 teeth (fig. 43A, C, E; plate 8)   *D. subobscura* p. 98
    Palps with 1 strong terminal bristle and
    1 subterminal bristle sometimes almost as long
    as the terminal one; upper tarsal comb with 7 to
    10 teeth, lower with 8 to 10 teeth; external
    process bluntly pointed and short; clasper very
    rounded and cup-shaped, containing a clasper
    comb obviously longer than wide with 7 to 10
    teeth (fig. 43B, D, F)   *D. ambigua* p. 93

23 Palps with 2 equally long bristles;
    mesonotum with 2 unclear longitudinal stripes;
    abdominal tergites very dark brown [black]
    without lateral yellow areas; ovipositor plate

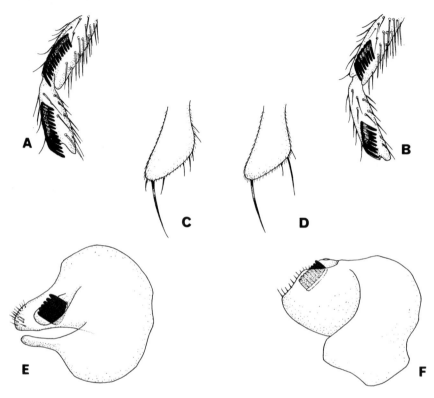

Fig. 43 *Drosophila subobscura*; **A** tarsal combs of male; **C** maxillary palp; **E** right genital arch and clasper of male. *Drosophila ambigua*; **B** tarsal combs of male; **D** maxillary palp; **F** right genital arch and clasper of male.

narrow and very pointed with conspicuous
bristles (fig. 44A, D, G)   *D. tristis* p. 99
Most often only 1 bristle on the palps;
if 2 bristles, then the subterminal bristle *less*
than half as long as the terminal one; if more
than half as long, lateral areas of the abdominal
tergites show yellow areas (fig. 44B, C, H, I)   24

24 Palps with 1 strong terminal bristle often with
a considerably smaller subterminal bristle;
mesonotum with distinct, longitudinal stripes;

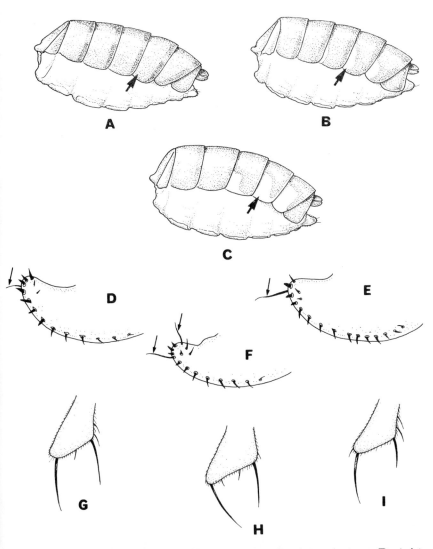

Fig. 44 *Drosophila tristis*; **A** female abdomen, lateral view; **D** right ovipositor guide; **G** maxillary palp.
Drosophila obscura; **B** female abdomen, lateral view; **E** right ovipositor guide; **I** maxillary palp.
Drosophila subsilvestris; **C** female abdomen, lateral view; **F** right ovipositor guide; **H** maxillary palp.

4th, 5th, and 6th abdominal tergites laterally with
a more or less distinct yellow area; occasionally
these are faint, and rarely they are missing
entirely; ovipositor plate with 1 long bristle
(fig. 44B, E, I; plate 8)   *D. obscura* p. 97
Mesonotum without pattern; if a trace of
mesonotal colour pattern *is* apparent then
abdominal tergites are without any light areas
laterally and the bristles on the ovipositor plate
are very short   25

25 Palps with 1 strong terminal bristle and
1 weak subterminal one; 4th, 5th, and 6th
abdominal tergites laterally with a distinct
yellow area; sometimes these areas are visible
also on the 3rd tergite; ovipositor plate with
2 long divergent bristles (fig. 44C, F, H)   *D. subsilvestris* p. 98
No yellow areas on abdominal tergites; only
1 ovipositor bristle and this is short   26

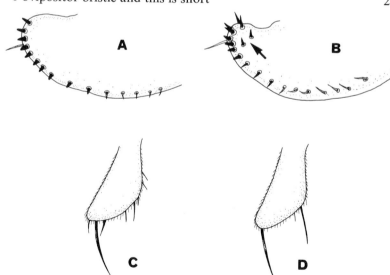

Fig. 45 *Drosophila subobscura*; **A** right ovipositor guide; **C** maxillary
palp.
Drosophila ambigua; **B** right ovipositor guide; **D** maxillary
palp.

26 Palps with only 1 conspicuous bristle;
teeth of ovipositor plate short and stout and
all of the same size; ovipositor bristle only about
twice as long as teeth (fig. 45A, C; plate 8)   *D. subobscura* p. 98
Palps with 1 strong terminal bristle, and
1 subterminal bristle sometimes almost as long
as terminal one; the 4 to 5 most terminal teeth
on the ovipositor plate longer and stouter than
the rest; ovipositor bristle about 3 times as long
as longest teeth (fig. 45B, D)   *D. ambigua* p. 93

Distribution and abundance of species described

D. ambigua Pomini 1940
Widely distributed in Europe (table on p. 100). However, it does not
appear to be very abundant, comprising at most 2% of the *Drosophila*
species coming to traps. Although it is recorded as a typical wild species
in Spain, in Britain and the rest of northern Europe it appears to be
principally associated with man. Thus, it has been recorded from farms,
gardens, orchards, greenhouses, public houses, and fruit stores. Less
frequently is has been found in woodlands and small copses away from
built-up areas. It has been trapped in Britain as early as May but it is
not taken in any numbers until June and July. Basden (1954a) suggests
that its late appearance, coupled with its high frequency in man-made
habitats, indicates that it is not a native, and that it does not overwinter
in Britain, but is introduced anew each year.

D. andalusiaca Strobl 1906 (previously called *D. forcipata* Collin 1952)
Appears to be rare, although it is locally abundant in certain habitats.
Basden (1954a) took only 2 males and 5 females, all in the Edinburgh
district. However, when he swept a net over a stream containing
watercress (*Nasturtium microphyllum*) for an hour he gathered 39 males
and 36 females. Thus, as with *D. fenestrarum*, its lack of response to
conventional bait can give a false impression of rarity. Beardmore (1967)
obtained many specimens near *Rumex* species and from rotting cucum-
bers. He suggests that the latter may be a better attractant in traps. The
species is found in numbers only in autumn.

D. busckii Coquillet 1901
A widely distributed and cosmopolitan species. In Britain it lives mainly
indoors and has been recorded from chicken-coops, greenhouses, and
fruit stores. In the survey carried out in southern England by Dyson-
Hudson (1954), 72% of all specimens collected were taken in or around

farms. Shaw (1968) records adults, larvae, and pupae on the open face of a covered pit of pea-silage, along with *D. hydei*. Several workers have found it breeding in large numbers among decaying plant matter and there is a record of 2 adult flies having been reared from cow-dung (Laurence, 1953). Workers in Switzerland observed large numbers of adults emerging from mushrooms, mainly species of *Russula*. It seems to be most abundant in Britain late in the year, particularly in August.

D. cameraria Haliday 1847 (erroneously called *D. pallida* Zetterstedt 1847)
Although this species has been recorded from many European countries it is seldom taken in large numbers. This is probably because it is reluctant to come to conventional *Drosophila* traps baited with fermenting fruit or *Drosophila* medium. It has been found near toadstools and on bracket-fungi on elm and oak. Outside the fungus season it can occasionally be trapped with conventional baits. It occurs almost exclusively in woodlands, small tree groups, and well-wooded gardens. July and August are the best months to look for it.

D. confusa Staeger 1844 (previously called *D. vibrissina* Duda 1924)
Appears to be quite widely distributed in Europe (table on p. 100) but never occurs in large numbers. Dyson-Hudson (1954) collected only 23 specimens, all but 1 from woodland habitats. Several specimens in the collection at the British Museum (Natural History) have been reared from fungi; indeed, *D. confusa* may well be a fungus-feeder. It appears to be most frequent in June.

D. deflexa Duda 1924
Basden (1954a) came to the conclusion that this species was uncommon in Scotland, though locally frequent. However, in the southern part of England it appears to comprise as much as 5% of the *Drosophila* fauna (Dyson-Hudson, 1954). It avoids close association with man and is found only in areas with trees or bushes. In the Dyson-Hudson survey, 99% of the 1496 specimens of *D. deflexa* collected were taken in such areas. This species is most frequent in July. Several observations indicate that it may be more numerous than suggested by the numbers coming to fruit-bait. Frydenberg (1956) records that at one site, 30 *Drosophila* were taken on an oak, attracted there by a mass of highly fermented sap. Of these, 28 were *D. deflexa*. Nearby, 88 flies were trapped on banana bait, but only 3 were *D. deflexa*.

D. fenestrarum Fallén 1823
Does not readily come to traps baited with fermenting fruit. Basden

(1954a) did not take a single specimen in traps, yet when he swept a net over a stream containing watercress (*Nasturtium microphyllum*) for an hour he obtained 19 specimens. This indifference towards the more usual types of bait would explain the small number of specimens recorded in the past. Frydenberg (1956) suggested that this species has no great preference for thickly wooded habitats. Almost all his specimens were taken in small groups of trees surrounded by open ground.

D. funebris (Fabricius 1787)

This is practically a world-wide species. The adults are found in or around buildings more often than they are in fields or woods. Dyson-Hudson (1954) took more than 50% of this species in or around farm buildings. However, it cannot be classified exclusively as an indoor species because it has also been found in gardens and orchards, in woodland of various kinds, and in more open habitats. Natural baits include toadstools and bracket-fungi collected together and allowed to decay. *D. funebris* can be caught at a wider variety of bait than any other species: it finds decaying material just as attractive as fermenting substances. Gordon (1942) records them from elm sap in late summer. They can be trapped all the year round but are not found outdoors until March. From then on they are found in increasing numbers until August.

D. helvetica Burla 1948

Appears to be an uncommon species in Britain. Basden (1954a) has no record of it in Scotland, but has found it frequently in southern Ireland (personal communication), and Dyson-Hudson (1954) took only 271 specimens, of which about 73% were found among trees and the remainder near farms. According to Burla (1951) and Hadorn et al. (1952), *D. helvetica* prefers immature woodland with a high humidity. This species is most frequent in June and July.

D. histrio Meigen 1830

Widely distributed in Europe (table on p. 100) but never very common. Dyson-Hudson (1954) collected only 46 specimens. Most of these were taken in woodland habitats, although specimens were collected from gardens and a farm also. *D. histrio* is most frequently encountered in June, although adults have been taken as early as April and as late as August.

D. hydei Sturtevant 1921

Closely associated with man in all parts of the world. Dyson-Hudson (1954) found it particularly common in gardens and orchards, where

85% of all the *D. hydei* specimens were taken; only about 11% were taken from woodland habitats. Shaw (1968) records adults, larvae, and pupae on the open face of a covered pit of pea-silage. Farmyards and decaying plants also attract this species. Sobels *et al.* (1954) suggest a preference for open habitats. For example, on the open slope of a hill with scattered groups of trees and bushes, *D. hydei* was the dominant species. It is most abundant from June until August. In Dyson-Hudson's survey (1954) 80% of all the specimens collected were taken during this period.

D. immigrans Sturtevant 1921

A cosmopolitan species. Burla (1951), in his investigation of the Swiss fauna, found very few *D. immigrans* north of the Alps. He concluded that it was restricted to the Mediterranean region. On the other hand, Sobels *et al.* (1954) found it quite common in the Netherlands. The highest frequencies occurred in or near woodland and in gardens. I have taken *D. immigrans* in Britain mainly from gardens and other cultivated areas. Basden (1954*a*) found it only in fruit stores and banana warehouses in Edinburgh. It can be trapped from May until December although, like most *Drosophila* species, it is most plentiful in autumn.

D. kuntzei Duda 1924

Basden (1954*a*) did not find this species in his Scottish collections. However, Dyson-Hudson (1954) records that it occurs with low frequency in all areas of southern England. In her survey it appeared to be almost exclusively a woodland species (97% of all specimens collected). Furthermore, 20% of these specimens were taken near water, suggesting a preference for damp situations. Adults can be caught from April until November, but the numbers fall off markedly by October. June seems to be the peak month.

D. littoralis Meigen 1830

A widely distributed species but never very abundant. It is often associated with fresh water. This was originally noted by Burla in 1951. Open land near small ponds, forests on the shores of lakes, and marshy or boggy woodland were all found to be suitable habitats. Rasmuson and Johansson (1969) note that in a trap situated at a riverbank *D. littoralis* was the dominant species, often giving up to a hundred individuals in a catch. They have also been reared off sap taken from the stump of a sycamore tree (Basden 1954*a*). The sap was collected at the end of August 1952 and larvae present in it produced 4 adults in the following September.

D. melanogaster Meigen 1830 (previously called *D. ampelophila* Loew 1862)
A cosmopolitan species recorded from all the European countries recently investigated. It is associated with man and occurs mostly near houses, often being especially numerous in gardens and orchards. In such places it is commonly found at bruised fruit such as apples, raspberries, and strawberries. Basden (1954*a*) found many in fruit stores and banana warehouses in Edinburgh in autumn. *D. melanogaster* can also be trapped in many natural habitats—mainly woodland—but here numbers are usually fairly low. It can be trapped outdoors from May to October, though the majority of flies occur in June and July. Indoors it can be trapped for most of the year.

D. obscura Fallén 1823 (previously called *D. obscuroides* Pomini 1940)
An extremely abundant and widespread species usually surpassed in numbers only by *D. subobscura* in most areas. It has been trapped in almost all types of outdoor habitat except for moorland and open coast. However, it is most abundant in woodland: in the survey conducted by Dyson-Hudson (1954), 72% of all *D. obscura* taken were from woodland habitats, 14% were from habitats associated with man, and a further 14% were from open habitats. There is some evidence that *D. obscura* reaches it highest frequencies in urban areas (see Chapter 3). Adult flies have been recorded from sap of elm, oak, lime, birch, alder, beech, willow, and sycamore. Gordon (1942) noted that a number of *Drosophila* eggs and larvae were found in the yeasty exudate uncovered when bark was removed from an elm tree. These were transferred to standard culture medium and from the 150 larvae collected, 20 male and 29 female *D. obscura* (but no other species) emerged. From December to March few flies are obtained. In April, however, numbers begin to build up and reach a peak from June to August, declining thereafter.

D. phalerata Meigen 1830
Widely distributed in European countries and very common in wooded areas. It seems to avoid open places and keeps away from man. Dyson-Hudson (1954) took 87% of all her specimens of *D. phalerata* in woodland of some kind. Mature deciduous and mixed woodland are the preferred habitats but it can also be taken in coniferous woodland and scrub. Its main breeding and feeding sites are thought to be toadstools although the flies select a wide range of fungi. For example, workers in Switzerland obtained 971 specimens from a total of 30 fungus species. Eggs and larvae can be found in healthy toadstools, and adults are obtained in large numbers by collecting fungi and rearing the larvae

from them in the laboratory. Adults come to traps from April to September although, as might be expected of a fungus-feeder, there is a large increase in numbers in autumn when toadstools are numerous.

D. simulans Sturtevant 1921

In northern Europe it occurs outdoors only rarely. Almost all the specimens taken by Basden (1954a) in Scotland were caught in fruit stores and banana warehouses in Edinburgh. Small numbers can usually be trapped in gardens and orchards during the warmer part of the year. In the Catalan region of Spain, although it was found in almost every part of the area investigated, it reached its highest frequency in urban areas (Monclus, 1964).

D. subobscura Collin 1936

This species is known from all European countries recently investigated. The adult can be found in almost every type of habitat: woodland, open arable country, gardens, orchards, sea-coast, moorland, and even the interior of buildings. However, it reaches its highest frequency in the neighbourhood of trees. Gordon (1942) found adults feeding on yeasty sap exudates of elm trees in Aberdeen. Basden (1954a) states that both sexes were commonly found on the bleeding stumps of sycamore and willow. There are also records of *D. subobscura* emerging from oak galls. Toadstools removed to the laboratory and kept in an insectory will quite often prove to be breeding material, too, and flies can be taken in the wild in the vicinity of healthy fungi. During the day there appear to be two peaks of activity: one between 0900 hr and 1100 hr and another between 1600 hr and 1800 hr (see Chapter 3). According to Basden (1954a), numbers reach their peak in Autumn and early winter. Dyson-Hudson (1954), working in southern England, obtained the highest catches between June and August. Basden records that adults were trapped in woods in the Edinburgh district in every month of the year.

D. subsilvestris Hardy and Kaneshiro 1968 (previously called D. silvestris Basden 1954)

The distribution of this species in Europe suggests that it prefers a north European climate. It is restricted to woodland habitats. (Dyson-Hudson, in 1954, took 99% of all specimens in such sites.) In these habitats it appears to be very common, often constituting 10% of the total *Drosophila* population. In some months (June and September) its numbers may even surpass those of *D. subobscura* and *D. obscura*. The adults are not in evidence until the latter part of April but they remain on the wing until November. There appear to be two peaks of abundance: one in June and July and another in September and October, with a clear decline in between. According to Basden (1954*a*), the adults are less active in winter than those of *D. subobscura, D. obscura,* and *D. tristis.*

D. transversa Fallén 1823

This is a fungus-feeder. The population numbers always seem to be very low and Basden (1954*a*) never caught adults before June. It prefers open localities with scattered trees and bushes. In Dyson-Hudson's 1954 survey, 62% of her specimens were taken in habitats which loosely could be called "tree groups", but only 15% were taken in woods. A point of interest was that 64% of the flies were taken in habitats close to ponds, streams, and fenland.

D. tristis Fallén 1823

Although widely distributed in Europe it usually represents only a small percentage of the total *Drosophila* fauna. An exception to this is a population reported by Sobels *et al.* (1954) in deciduous woodland in which *D. tristis* was the dominant species and comprised 54% of the total specimens collected. It appears to be confined to woodlands of various kinds. Dyson-Hudson (1954) took 79% of her total catch for this species in woods. *D. tristis* can occasionally be found in gardens and orchards but seems never to enter buildings. It has been recorded feeding on exuding tree sap. Adults are on the wing from April until November but are most common in July and August.

European distribution records for 22 species of *Drosophila*

	Denmark	Eire	Finland	France	Germany	Great Britain	Greece	Italy	Netherlands	Portugal	Spain	Sweden	Switzerland	recorded additionally in:
D. ambigua	x	x		x	x	x	x	x	x	x	x	x	x	Austria
D. andalusiaca	x	x		x	x	x			x		x	x	x	Yugoslavia
*D. busckii**	x	x	x	x	x	x			x	x	x	x	x	Norway
D. cameraria			x	x	x	x			x		x		x	
D. confusa	x	x	x	x	x	x	x	x	x	x	x	x	x	
D. deflexa		x		x	x	x			x				x	
D. fenestrarum	x	x	x	x	x	x	x	x	x	x	x	x	x	Hungary, Poland
*D. funebris**	x	x	x	x	x	x	x		x	x	x		x	Hungary, Poland
D. helvetica	x	x	x	x	x	x			x		x	x	x	Iceland, Russia
D. histrio				x	x	x		x	x		x		x	
*D. hydei**	x	x	x	x	x	x	x		x	x	x	x	x	
*D. immigrans**	x	x	x	x	x	x		x	x		x		x	
D. kuntzei				x	x	x			x		x		x	
D. littoralis	x	x	x	x	x	x	x	x	x	x	x	x	x	Austria, Russia
*D. melanogaster**	x	x	x	x	x	x	x		x	x	x	x	x	Hungary
D. obscura	x	x	x	x	x	x	x	x	x	x	x	x	x	Austria, Russia
D. phalerata	x	x	x	x	x	x	x		x	x	x	x	x	Austria
D. subsilvestris	x	x	x		x	x			x		x		x	Russia
*D. simulans**	x	x	x	x		x	x		x	x	x		x	
D. subobscura	x	x	x	x	x	x	x	x	x	x	x	x	x	Belgium
D. transversa	x	x	x	x	x	x		x	x	x	x	x	x	
D. tristis	x	x	x	x	x	x		x	x		x	x	x	

*not native to Europe

100

Check list of British species of *Drosophila*

sub-genus *Drosophila*

Quinaria group
Drosophila kuntzei Duda
Drosophila limbata von Roser
Drosophila phalerata Meigen
Drosophila transversa Fallén

Virilis group
Drosophila littoralis Meigen
Drosophila unimaculata Strobl

Testacea group
Drosophila testacea von Roser

Funebris group
Drosophila funebris (Fabricius)

Repleta group
Drosophila hydei Sturtevant
Drosophila repleta Wollaston

Polychaeta group
Drosophila polychaeta Patterson
and Wheeler

Immigrans group
Drosophila immigrans Sturtevant

Melanderi group
Drosophila cameraria Haliday

Unassigned species
Drosophila confusa Staeger
(= *D. vibrissina* Duda)
Drosophila histrio Meigen
Drosophila iri Burla
Drosophila picta Zetterstedt

sub-genus *Lordiphosa*

Fenestrarum group
Drosophila acuminata Collin
Drosophila andalusiaca Strobl
(= *D. forcipata* Collin)
Drosophila fenestrarum Fallén

sub-genus *Pholadoris*

Drosophila deflexa Duda
Drosophila rufifrons Loew

sub-genus *Dorsilopha*

Drosophila busckii Coquillet

sub-genus *Sophophora*

Melanogaster group
Drosophila ananassae Doleschall
Drosophila melanogaster Meigen
(= *D. ampelophila* Loew)
Drosophila simulans Sturtevant

Obscura group
Drosophila ambigua Pomini
Drosophila helvetica Burla
Drosophila obscura Fallén
(= *D. obscuroides* Pomini)
Drosophila subsilvestris Hardy
and Kaneshiro (= *D. silvestris*
Basden)
Drosophila subobscura Collin
Drosophila tristis Fallén

The sub-genera *Hirtodrosophila, Phloridosa, Sordophila,* and *Siphlodora.* as well as some tropical ones contain no British representatives.

8 Techniques of study

When putting into practice the techniques outlined in this chapter, the successful application of most of the methods depends on a thorough knowledge of the biology of *Drosophila*. For example, when trapping flies in the wild it is necessary not only to know about suitable traps and baits, but also about the distribution in time and space of the various species you are looking for (see Chapter 3).

Methods of collecting

In the main, collecting involves attracting fruit flies to bait. Fermenting fruit of various kinds is particularly effective, especially apple or banana. Treat the apples in the following way. Chop them into small pieces and cook them for several minutes to kill any insects already present in the fruit. (Basden [1954*a*] notes that *D. funebris, D. melanogaster,* and *D. subobscura* were reared directly from bruised apples bought from a fruit shop.) Place the fruit in a fly-proof container for about a week to allow fermentation to take place. Keep the container at a temperature of 25° C and, each day, stir the fruit to aid fermentation and prevent the growth of moulds. After about 7 days the apple-bait is ready for use and will retain its attractiveness for up to 4 weeks. Although apple has been found to be most effective, you can try other prepared fruit baits. Furthermore, decaying vegetables, meat, and fungi, while not attracting the same large number and variety of flies as fermenting fruits, may prove more effective for collecting certain species.

Bait can be exposed in a variety of ways; three methods will be described here.

1 If time is very short the best method is simply to spread the bait in a thin layer over the ground or on a convenient branch of a tree. Any *Drosophila* can then be collected using a net and/or a suction-bottle, which can be bought from an entomological dealer, or made. There is a variety of suction-bottles (called *pooters*) but the operating principle behind all of them is the same. A typical pooter is shown in Figure 46. The flies are drawn into the glass bottle A *via* the tube B by sucking through the second

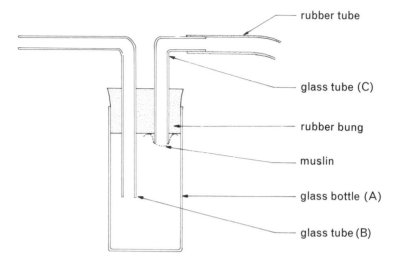

rubber tube

glass tube (C)

rubber bung

muslin

glass bottle (A)

glass tube (B)

Fig. 46 A pooter.

Fig. 47 *Drosophila* traps; *left*, cup trap; *right*, bottle trap.

tube C. They are prevented from entering tube C by a small piece of muslin placed over the end protruding into the glass bottle. The captured flies can then be etherized and put into breeding tubes containing food or, alternatively, into collecting bottles containing 70% alcohol with 5% glycerine. *If living specimens are required take great care not to over-etherize the flies.*

2   If there is more time and a wider area is sampled, the bait is better placed in small, plastic drinking cups. These are quite cheap and are very light to carry in the field. They may be placed at ground-level or hung from a tree at an angle of 45° using string tied through holes in the lip and side of the cup (fig. 47, left).

As you can see in Figure 48, the flies at the bait are removed by a glass funnel and tube placed over the open end of the cup. Since the flies are attracted by light—they show *positive phototaxis*—they fly upwards into the collecting tube. These cups can be put out in the late evening and visited several times next day. They are especially effective if *Drosophila* are plentiful and if you can inspect the cups at a time of day when the flies are active.

3   If, however, the numbers of *Drosophila* being attracted are small, it is

Fig. 48 Removing flies from a cup trap in the wild.

far better to place the bait in some kind of trap which allows flies to enter but not to leave. Such traps can then be left out for several days before being collected and removed to the laboratory. Similarly, they are useful if you can inspect the traps only at a time of day when the flies would normally be inactive. A *Drosophila* trap is shown in Figure 47, right. Its construction is simple. The main body of the trap is a 250-cm^3 milk bottle. In the top is a cork bored with two holes, each taking a 5-cm length of tubing of 10 mm internal diameter. The two tubes are level with the top surface of the cork, but project a little way into the trap, thus making it easy for flies to enter but difficult for them to leave.

Since this type of trap is usually left outside for several days, some kind of shelter should be placed over it. In very exposed sunny areas it is advisable to place a cardboard collar round the bottle-neck to shade the trapped flies. The flies can be removed from the trap in the field, or in the laboratory by means of the suction tube and the cork top replaced with cotton-wool. Any eggs laid by the captured flies can be reared and the F_1 adults examined.

With both the bottle trap and plastic cups it is often necessary, with very liquid bait, to lay a piece of paper tissue or towelling on its surface to provide the flies with a firm foothold.

For reference it is advisable to keep the specimens collected during a survey or project in a convenient form in the laboratory. I have found that collections are best kept in small glass tubes containing 70% alcohol and glycerine (5%). These can be stored in blocks of plywood bored with appropriately sized holes. Each collection should be labelled and a record kept of all relevant information concerning the site of capture. For this purpose, record sheets prepared in advance and filled in at the site of collection are most convenient. An example of such a record sheet is shown in Figure 49.

Laboratory cultures

Most of the 22 species of British *Drosophila* described in this book can be maintained in the laboratory. For experimental purposes, adequate cultures can be obtained by placing about 15 mm of food medium into a 75 mm glass tube. A small piece of paper tissue pushed into the medium before it becomes solid will absorb any excess moisture and also provide a pupation site for the larvae. A cotton-wool bung prevents the flies escaping and contamination from other flies. All cultures should be labelled with the name of the person setting up the culture, the date, any relevant information concerning the parental flies, and the purpose of the experiments. If larger populations are required, 250-cm^3 bottles set up in a similar way to the glass tubes are useful.

name of collector	B. Shorrocks		collection number	56 / Nc
date	6 September 1968	time	17. 00	

description of site

Dipton Wood, Northumberland (550 ft. above sea level). Open coniferous wood with very little ground vegetation. trap placed 20 metres from edge of wood.

map reference 35 976615

weather

Warm and sunny. no cloud. no rain. no wind (Univ. Geog. Dept. met. reading - $7\frac{1}{2}$ hrs. sunshine. max. temp. 21°C. 4.5mm. rain previous day)

temperature	humidity	light intensity
15.5°C (air)	65%	no reading taken.

flora

(see Tyne file. collection site 11)

type of bait and trap	fermenting apple cup trap		height of trap above ground	ground level .

species collected	male	female	additional information
phalerata	-	2	trap had been disturbed - lying on its side .
silvestris	-	2	
subobscura	2	6	
obscura	-	4	
immigrans	-	1	
Total	2	15	

Fig. 49 A record sheet.

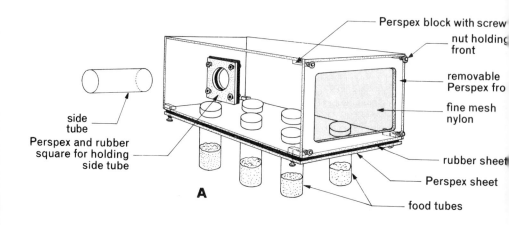

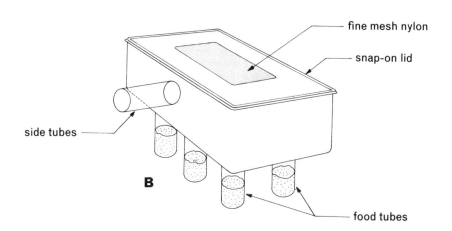

Fig. 50 Population cages; **A** complex cage made from Perspex; **B** simple cage made from a polythene sandwich box.

For certain ecological and evolutionary studies, larger containers or population cages, into which a number of food tubes can be inserted and replaced at regular intervals, are best. Stock cultures of wild species are very easily maintained in these cages. They can be made from polythene sandwich boxes (fig. 50B) with holes cut in the bottom (and sides if required) of just sufficient diameter to take 75 mm × 25 mm food tubes. These tubes are prepared in the way described in the previous paragraph, except that they have no cotton-wool bungs. The food tubes are held in place by the walls of the polythene cage. A larger hole can be cut in the lid and covered with fine-mesh muslin or nylon for aeration. Figure 50A is a more elaborate population cage made from Perspex. Here, the food tubes are held by a rubber sheet trapped between the bottom of the cage and another sheet of Perspex. The six holes in the rubber sheet are slightly smaller than the ones in the Perspex on either side. The protruding ring of rubber acts as a gripping collar, holding the food tubes in position. In this particular cage, a tube is also insertable at the side. Apart from providing an additional food tube for sampling fly populations it also means that two cages can be joined together.

Food media Many culture media are now used in the laboratory breeding of *Drosophila*. Only two will be described here. The first is a standard maize meal medium which has proved very successful in laboratories all over the world. The following ingredients are required:

cornmeal (maize meal)	13·5 g
molasses	18·0 g
agar	2·0 g
nipagin	0·3 g
water	100·0 cm³

These quantities will make enough food for twenty-four 75 mm × 25 mm glass tubes or four 250-cm³ milk bottles. Pour about two-thirds of the water into a pan (a double porridge pan is best because it prevents the food from burning). Add the cornmeal and molasses to the cold water, bring the mixture to the boil, and cook it for several minutes. Then add the agar and nipagin to the remaining cold water and mix them thoroughly. When the cornmeal and molasses mixture is ready, tip in the agar/nipagin mix and cook the whole for a few minutes more, stirring constantly. The medium is ready for use when it is still thin enough to pour easily but will gel firmly on cooling. Two important points to remember are that all the ingredients should be carefully mixed with cold water first to prevent lumps forming, and that the food should be stirred constantly. The ingredient nipagin is a mould inhibitor and can be obtained in 100-g packets from Nipa Laboratories Limited, Treforest Industrial Estate, South Wales. The day before flies are to be put into the containers, a drop of a thick suspension of fresh yeast-cake in water should be added. (Dry yeast may be used instead of fresh yeast but results are not usually so good.)

The second food medium is my own and can be prepared very quickly. Fungus-feeding *Drosophila*, which are often difficult to breed on the standard laboratory medium, have been found to perform well on this food. The following ingredients are required:

instant breakfast cereal (rolled oats)	5·0 g
brown sugar	5·0 g
dried yeast	12·0 g
agar	3·0 g
nipagin	0·3 g
water	100·0 cm^3

Mix the breakfast cereal, brown sugar, and dried yeast with about two-thirds of the cold water in a pan. Add the agar and nipagin to the remaining one-third cold water. Bring the mixture in the pan to the boil and cook it for just a few minutes until the yeast has been killed. Once the agar and mould inhibitor have been added, the whole should be cooked for about one minute (when the medium will gel firmly on cooling, but will pour easily while hot). No yeast suspension need be added to the culture bottles before the adult flies are introduced.

The standard medium can be stored in bottles for up to two months. After preparing and pouring the food into the culture bottles they are capped with kitchen foil and sterilized for 20 min at 120° C. After any reasonable storage period at 18° C it is necessary only to add the live yeast suspension and the bottles are ready for use.

D. subsilvestris has been kept in the laboratory by Basden (1954*a*) using the following technique. The animals are maintained in a population cage where one of the food tubes contains 4 cm^3 of honey with a wisp of dry cotton-wool on top on which the flies can stand when the honey *deliquesces* (liquefies). Another tube should contain cotton-wool soaked in water for drinking. Wide tubes intended for larval development contain prepared apple and *Drosophila* medium in equal proportions, but unmixed. The wet cotton-wool will need moistening occasionally and a new food tube for egg-laying should be added and a spent one removed each week.

D. deflexa can be maintained in a similar way, but without the honey. The maintenance of this species is further complicated by the fact that the larvae go into diapause from about October to June.

In the early days of *Drosophila* culturing, moulds presented a great problem in the laboratory. The use of mould inhibitors of various kinds has now greatly reduced this source of trouble. However, cultures may still become infected occasionally. If you discover contaminated cultures, transfer the flies immediately to freshly made up bottles or cages, and destroy the old cultures. Perhaps more serious than mould infections are mite infections (fig. 51). Several species of mites have been found in *Drosophila* cultures and in almost all cases a heavy infestation generally coincides with a poor yield of flies. The mites use up food and tend to dry out the surface of the medium. It is also possible that they feed on *Drosophila* eggs. Some of the mites attach themselves to adult flies in such numbers that they hamper movement. Others puncture the integument and suck tissue fluids. As with mould infections, the infested culture should be destroyed after removing a number of adult flies for breeding.

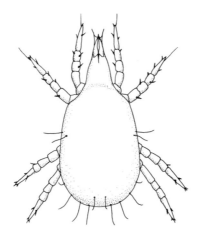

Fig. 51 A laboratory mite, *Histiostoma* species.

These should be transferred to freshly made up culture bottles and this process continued every few days until the stock is clear of infection.

Culture bottles and population cages can be kept free from mite infection by spraying them with a 2% solution of benzyl-benzoate in alcohol. This chemical does not harm *Drosophila*. Shelves, trays, and tables can be wiped over with an aqueous solution of 1:100 phenol or lysol, both of which are effective fungicides and germicides. By far the most important point to remember in breeding *Drosophila* is that a clean laboratory prevents most of these infections. All old culture bottles should be sterilized by heating them and never be allowed to remain uncleaned in the laboratory for long periods. Cotton-wool, paper tissues, and the glass tubes or bottles themselves should be heat-sterilized before use. Finally, stocks of flies received from other laboratories, and individual flies caught in the wild, must be examined thoroughly before being brought among existing cultures.

Handling flies   To examine flies and transfer them to culture bottles, they must be anaesthetized. Ether is normally used; carbon dioxide less frequently. Etherizers are easy to make and two examples are illustrated in Figure 52.

Fig. 52 Equipment for handling *Drosophila* in the laboratory.

The first (second from left) is simply a glass jar with a ground-glass lid. Inside the hollow of this lid is fastened a wad of cotton-wool. Ether is introduced onto this from a dropping-bottle. The flies are shaken from the culture bottle into the etherizer and the lid replaced. Immediately movement ceases the flies must be removed.

The second type of etherizer (third from left) is more complex but much more efficient. Here we have a plastic tube with a clip-on plastic lid. The glass bottle into which the plastic tube is inserted has a neck diameter of just sufficient size to allow the tube, but not its clip-on lid, to pass through. The tube is suspended, therefore, inside the bottle. A hole in the lid of the tube receives the shortened narrow end of a plastic funnel. Small holes are made in the bottom of the plastic tube with a red-hot needle so that ether fumes can pass through but not flies. In the bottom of the glass bottle is a wad of cotton-wool dampened with a few drops of ether. Flies shaken from the culture bottle into the plastic tube *via* the funnel are anaesthetized by the ether fumes. The ether-soaked cotton-wool must not come into contact with the bottom of the plastic tube because the ether will dissolve it.

The fumes of ether will eventually have a detrimental effect upon the plastic but this part of the etherizer is easily replaced. For this reason the tube should not be left inside the bottle for long. Glass tubing with muslin sealed at the bottom in place of the plastic tube would overcome this.

If the flies begin to recover before you have finished examining them they can be re-etherized using the apparatus shown in Figure 52 (fourth from left). This is simply a petri dish with a small wad of cotton-wool covered with muslin fastened to its under-surface. If a small amount of ether is added to the cotton-wool pad, flies covered with the inverted dish will remain anaesthetized. The cork attached to the top acts as a handle. Flies that have been examined and are no longer required can be dropped into a "morgue". This is a glass bottle containing alcohol or oil.

Examination is best carried out on a white tile or glass surface using a camel-hair brush or a fine pair of forceps for manipulation, taking care not to over-etherize the flies. Flies accidentally killed in this way are easy to recognize because they have their wings extended at an angle of 45° above the body. Two final points to remember when setting up cultures are first, remove any excess moisture from the walls of the container before the flies are introduced, and second, if the flies have not yet recovered from etherization they should not be dropped directly onto the food surface. Instead, the bottles or tubes should be laid on their sides with the flies on the dry glass surface until they have recovered.

Preparing salivary-gland chromosomes

Salivary-gland chromosomes can be seen clearly using a compound microscope with a × 10 objective and a × 10 ocular; a × 40 objective will make observation even easier. If you need to identify the details of the banding pattern you will need an oil-immersion objective. A green filter placed between the microscope and the light source will improve the degree of contrast in banding and so make identification of individual chromosomes and parts of chromosomes easier.

This section describes the preparation of salivary-gland chromosomes of *D. melanogaster*. The process is not a difficult one, but it requires a certain amount of practice before preparations of a good quality are produced consistently.

The first prerequisite is to culture large, well-fed larvae. Adult flies should be transferred every two days to freshly made-up containers. In this way only a few eggs will be laid in each tube or bottle and, consequently, the larvae will be uncrowded and well fed. To help this feeding process, a little fresh yeast can be added to the cultures when the parents are removed. At the same time the culture bottles should be removed to a cool place (about 16° C to 18° C) which will extend the larval period and, therefore, the time available for feeding.

Fully grown larvae are required, and if no paper tissue is placed in the cultures, the final instar larvae will climb up the sides of the bottle ready for pupation. From this position they can be picked up easily with a camel-hair brush.

Temporary preparations   In making temporary preparations of salivary-gland chromosomes, three solutions are required. These are:

a   Fixing solution: 1 part acetic acid to 3 parts absolute alcohol.

b   Staining solution: 2% orcein in 50% acetic acid. (Filter it well before use.)

c   Ringing solution: a saturated solution of gelatine in acetic acid.

In all cases the acetic acid is glacial acetic acid. If the staining solution proves too concentrated, dilute it with 50% acetic acid.

Dissect a third-instar larva in a drop of salt solution (7 parts of sodium chloride in 1000 parts of distilled water). Remove the salivary glands with the aid of two dissecting needles. Hold the larva firmly across the middle with one needle and place the other just behind the mouthparts. If you draw this second needle slowly forward, the mouthparts and attached structures, including the salivary glands, are pulled out of the larval body. The glands are semi-transparent and look like small, neat bunches of grapes (fig. 53). They should not be confused with the more opaque and less regular fat body, pieces of which may also be removed by the decapitating action of the needle. It is important that the glands are never allowed to dry out at any stage during the preparation.

Transfer the glands to a drop of solution *a* on a microscope slide. Leave them for a few minutes until they have turned white. Add 1 drop of solution *b* and leave them for about a further 10 minutes; experience will show whether a shorter or longer time is needed. (Chromosomes that have been fixed and stained for too long may be brittle, with poor differentiation between bands of different thicknesses; chromosomes that are under-stained may spread well but show indistinct banding.) Next, cover the glands with a coverslip and squash them by applying pressure immediately above the gland material. This is best done with the ball of the thumb, having first covered the slide with some filter-paper to absorb the excess stain that will be expelled. It is not possible to indicate the exact amount of

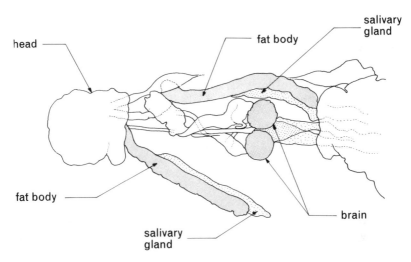

Fig. 53 Dissected third instar larva showing salivary glands.

pressure required to make a good squash. The chromosomes should be spread, but not broken; once again experience helps. To prevent evaporation of the fluid under the coverslip, a small amount of solution c run around the edge of the coverslip will protect the preparation for several months.

Permanent preparations If permanent preparations are required the coverslip must be removed after squashing and the chromosomes treated. First it is necessary to anchor the chromosomes to the slide. Slides to be used are first covered with a thin coat of albumen solution. This solution is prepared by separating the white of an egg and adding to it an equal quantity of glycerol and then 0·5 g of thymol (a preservative).

Place a drop of albumen solution at one end of the slide and spread it along the surface with the edge of another slide. Allow the film to dry before use.

The initial part of the preparation up to and including the squashing is the same as for the temporary preparation. After squashing, place the slides in a dish containing a layer of paper towelling covered with 95% alcohol. The lower edge of each coverslip must just be allowed to touch the alcohol. After about 12 hours the slides can be immersed in 95% alcohol, when the coverslips will fall off. (If they do not they can be carefully removed with the edge of a razor blade.) Then put the slides into absolute alcohol for two minutes, with a change of the alcohol after one minute. Finally, clear the preparation in xylene (xylol) and mount it in Canada balsam. The preparation can be mounted in euparal if you prefer it, in which case no clearing in xylene is necessary. At no time should the preparation be allowed to dry out.

Analysing results of breeding experiments

In many of the genetic experiments outlined in Chapter 4 and in the Investigations certain ratios of offspring are expected in particular generations. Often, the numbers of flies counted do not fit exactly the expected proportions and it becomes necessary to look at the results more closely. Departures from expected results may be caused by chance, by experimental conditions causing excessive mortality of the mutants, or even by a faulty hypothesis.

The chi-square (χ^2) test involves determining a predicted ratio, and then establishing how closely the observed data fit this ratio. Consider the following data obtained by inbreeding *Drosophila* of the genotype $+/vg\ +/e$:

phenotype	number
normal (+)	2834
vestigial (vg)	920
ebony (e)	951
vestigial and ebony (vge)	287
Total	4992

The χ^2 test for the data in this table is performed as follows:
1   The hypothesis: that the observed phenotypic ratio is consistent with a predicted ratio of 9/16 + :3/16 vg:3/16 e:1/16 vge.

2 The expected or predicted ratio: dividing the total number of 4992 by 16, the expected number of vge flies is calculated to be 312. The expected number of normal flies is, therefore, $9 \times 312 = 2808$, and the expected number of vg (and of e) is $3 \times 312 = 936$.

3 The deviations between observed and expected ratios and the calculation of χ^2:

$$\chi^2 = \sum(O - E)^2/E$$

where capital sigma ($\sum$) means "sum all terms like", O is the observed experimental figure, and E is the expected figure.

	+	vg	e	vge	total
observed (O)	2834	920	951	287	4992
expected (E)	2808	936	936	312	4992
O–E	+26	–16	+15	–25	
$(O-E)^2$	676	256	225	625	
$(O-E)^2/E$	0·24	0·27	0·24	2·00	
$\chi^2 =$	0·24 +	0·27 +	0·24 +	2·00 =	2·75

4 The probability value (P): tables of χ^2 values have been prepared from which it is possible to determine the probability that a certain value of χ^2 is either small enough to be regarded as a chance departure from zero, or large enough to indicate that something other than chance is operating. (Note: if the expected and observed values are the same then $\chi^2 = 0$. Therefore, the greater the difference between the two ratios, the larger will be the value of χ^2.) The relationship between χ^2 and P is determined by the *degrees of freedom* (df). If N is the total number of classes, the degrees of freedom usually equal $N-1$. In the above case, $N-1$ is $4-1 = 3$. For a value of $\chi^2 = 2·75$ with 3 degrees of freedom, written χ^2_3, the value of P obtained from the table lies between 0·50 and 0·30. What this means is that such a value of χ^2 could have been obtained between 50 and 30 times out of 100 purely by chance. There is, therefore, nothing significant about the deviation between observed and expected ratios here. Such a deviation could have occurred easily by chance and our data appear to satisfy a 9:3:3:1 ratio as predicted. A χ^2 value is regarded as being *significant* if the table of values indicate that it would occur less than 5 times out of 100 ($P \leqslant 0·05$) by chance. If $P \leqslant 0·01$, then the observed departure from the expected is said to be very significant. In both these cases we assume that chance alone does not account for the results. A table of χ^2/probability values appear overleaf.

probabilities

df	·95	·90	·70	·50	·30	·20	·10	·05	·01	·001
1	·004	·016	·15	·46	1·07	1·64	2·71	3·84	6·64	10·83
2	·10	·21	·71	1·39	2·41	3·22	4·61	5·99	9·21	13·82
3	·35	·58	1·42	2·37	3·67	4·64	6·25	7·82	11·35	16·27
4	·71	1·06	2·20	3·36	4·88	5·99	7·78	9·49	13·28	18·47
5	1·15	1·61	3·00	4·35	6·06	7·29	9·24	11·07	15·09	20·52
6	1·64	2·20	3·83	5·35	7·23	8·56	10·65	12·59	16·81	22·46
7	2·17	2·83	4·67	6·35	8·38	9·80	12·02	14·07	18·48	24·32
8	2·73	3·49	5·53	7·34	9·52	11·03	13·36	15·51	20·09	26·13
9	3·33	4·17	6·39	8·34	10·66	12·24	14·68	16·92	21·67	27·88
10	3·94	4·87	7·27	9·34	11·78	13·44	15·99	18·31	23·21	29·59
11	4·58	5·58	8·15	10·34	12·90	14·63	17·28	19·68	24·73	31·26
12	5·23	6·30	9·03	11·34	14·01	15·81	18·55	21·03	26·22	32·91
13	5·89	7·04	9·93	12·34	15·12	16·99	19·81	22·36	27·69	34·53
14	6·57	7·79	10·82	13·34	16·22	18·15	21·06	23·69	29·14	36·12
15	7·26	8·55	11·72	14·34	17·32	19·31	22·31	25·00	30·58	37·70
20	10·85	12·44	16·27	19·34	22·78	25·04	28·41	31·41	37·57	45·32
25	14·61	16·47	20·87	24·34	28·17	30·68	34·38	37·65	44·31	52·62
30	18·49	20·60	25·51	29·34	33·53	36·25	40·26	43·77	50·89	59·70
50	34·76	37·69	44·31	49·34	54·72	58·16	63·17	67·51	76·15	86·66

not significant   significant

118

INVESTIGATIONS

For most of the genetics experiments described here it is essential to use virgin females. Since females do not normally mate until about 12 hours after emergence, young newly emerged females should be collected and put into food tubes without males. The simplest procedure is to remove all the adults from a stock bottle and collect all the females emerging during the following few hours. Because of the emergence rhythm shown by most species, the best time to clear *all* unwanted adults is in the late afternoon or evening; thus, the virgin females can be collected the following morning.

In the chapter on genetics it was mentioned that differences in the mortality rates of the various mutant and normal flies often means that the observed genetic ratios are different from those expected. Because of this, overcrowding of the culture bottles should be avoided. In addition, several counts can be made of the emerging generation. After the first examination and classification the cultures are not destroyed. Instead, the bottles are left for a further 2 days and the freshly emerged flies counted. Both sets of data can be combined for the final comparison of expected and observed results. Counting of the progeny over a period of a few days also prevents the occurrence of odd ratios caused by different development times of some mutant strains.

A cross involving one mutant character (monohybrid cross)
In the following account ebony is the mutant used; a list of the other mutants that could be used is given in the Appendix.
Day 1 Cross a virgin ebony (e/e) female with a normal (+/+) male. If 75 mm × 25 mm tubes are being used, three pairs of flies are needed per tube. For experiments in milk bottles at least six pairs should be used. Cultures should be kept at approximately 25° C.
Day 6 The tunnelling of the larvae should be clearly visible in the cultures, and the parental flies should now be removed.
Day 14 Count and examine the F_1 progeny, all of which should have the normal phenotype. Mate an F_1 female with an F_1 male and set up fresh cultures using the same number of pairs as before. In this case, the females

119

need not be virgins and can be taken directly from the examined F_1 flies.

Day 20 Remove the F_2 parents.

Day 28 Count and examine the F_2 progeny. The complete results of the experiment can be entered in a table like the one shown here. Using the χ^2 test, see if your observed results fit the expected $3:1$ ratio.

name			date	
cross ebony (e/e) female × normal (+/+) male				
F_1 date counted	normal			
total				
F_2 date counted	normal	ebony	total	
observed numbers (O)				
expected numbers (E) 3:1				
$(O-E)^2/E$				
χ_1^2				

2 A back-cross

Day 1 Cross a virgin ebony (e/e) female with a male (+/e) from the F_1 of experiment 1.

Day 6 Remove the F_1 parents.

Day 14 Count and examine the back-cross generation. Enter your results in a table similar to the one shown at the top of page 121 and determine if the observed data fit the expected $1:1$ ratio.

3 A cross involving two mutant characters (dihybrid cross)

Day 1 Cross a virgin ebony/vestigial (e/e vg/vg) female with a normal (+/+ +/+) male or

Cross a virgin ebony (e/e +/+) female with a vestigial (+/+ vg/vg) male.

name		date	
cross ebony (e/e) female × normal (+/e) male			
back-cross progeny, date counted	normal	ebony	total
observed numbers (O)			
expected numbers (E) 1:1			
$(O-E)^2/E$			
χ^2_1			

name				date	
cross ebony/vestigial (e/e vg/vg) female × normal (+/+ +/+) male					
F_1 date counted	normal				
total					
F_2 date counted	normal	ebony	vestigial	ebony and vestigial	total
observed numbers (O)					
expected numbers (E) 9:3:3:1					
$(O-E)^2/E$					
χ^2_3					

Day 6 Remove the parents.
Day 14 Count and examine the F_1 generation, all of which should be normal in appearance. Mate an F_1 female with an F_1 male. The females need not be virgins.
Day 20 Remove the F_2 parents.
Day 28 Count and examine the F_2 progeny. Enter your results in a table similar to the one shown at the foot of page 121, and determine whether they fit the expected 9:3:3:1 ratio.

4 A cross involving a sex-linked character

Use two reciprocal crosses to give different results.
First cross: normal $(+/+)$ virgin female $\times$ white-eyed $(w/\leftarrow)$ male
Second cross: white-eyed (w/w) virgin female $\times$ normal $(+/\leftarrow)$ male.
Day 1 Set up the two crosses.
Day 6 Remove the parents.
Day 14 Count and examine the F_1 generation and set up new cultures using F_1 flies. The females need not be virgins.
Day 20 Remove these F_2 parents.
Day 28 Count and examine the F_2 generation and compare the results of the two reciprocal crosses.

name					date	
cross						
F_1 date counted	males		females			total
	normal	white-eyed	normal	white-eyed		
total						
F_2 date counted	males		females			total
	normal	white-eyed	normal	white-eyed		
total						

5 A three-point test-cross

Day 1 Cross a white/miniature/Bar (wmB/wmB) virgin female with a normal $(+ + +/\leftarrow)$ male.
Day 6 Remove the parents.

	+++ / Y	wmB / Y	+m+ / Y	w+B / Y	w++ / Y	+mB / Y	++B / Y	wm+ / Y	total
males	normal	white-eyed miniature Bar	miniature	white-eyed Bar	white-eyed	miniature Bar	Bar	white-eyed miniature	
date									
date									
total males									
	+++ / wmB	wmB / wmB	+m+ / wmB	w+B / wmB	w++ / wmB	+mB / wmB	++B / wmB	wm+ / wmB	total
females	semi-Bar	white-eyed miniature Bar	miniature semi-Bar	white-eyed Bar	white-eyed semi-Bar	miniature Bar	Bar	white-eyed miniature semi-Bar	
date									
date									
total females									
total males and females	A	B	C	D	E	F	G	H	

Day 14 Place some of the F_1 flies (males and females) into fresh food tubes. Virgin females are not required.
Day 20 Remove the F_2 parents.
Day 28 Count and classify the F_2 generation. Enter your results in a table like the one shown on page 123.
A and B are the parental classes. C and D are the double cross-over classes. The cross-over value between w and m is:

$$\frac{C+D+E+F}{total} \times 100 = 34\cdot6$$

The cross-over value between m and B is:

$$\frac{C+D+G+H}{total} \times 100 = 20\cdot9$$

1 Draw the chromosome map for these three loci, indicating their position relative to each other.
2 Calculate the coefficient of coincidence.

6 Analysis of linkage using χ^2

Day 1 Cross a white/miniature (wm/wm) virgin female with a normal $(++/ \text{—})$ male.
Day 6 Remove the parents.
Day 14 Place some of the F_1 flies (males and females) into fresh food tubes. Virgin females are not required.
Day 20 Remove the F_2 parents.
Day 28 Count and classify the F_2 generation. Enter your results in a table like this one:

$\frac{++}{}$ or $\frac{++}{wm}$	$\frac{wm}{}$ or $\frac{wm}{wm}$	$\frac{w+}{}$ or $\frac{w+}{wm}$	$\frac{+m}{}$ or $\frac{+m}{wm}$
normal	white-eyed miniature	white-eyed	miniature
a	b	c	d

Alternatively, the data obtained in experiment 5 can be used.
Considering only the w and m loci, the results of this experiment are grouped together so that data appropriate for experiment 6 are obtained. Thus, a = A+G; b = B+H; c = D+E; and d = C+F.
Testing for linkage If the two loci were assorting independently—that is, on different chromosomes, the ratio of non-cross-over classes (a+b) to cross-over classes (c+d) would be equivalent to a 1:1 ratio. We can

examine this using the χ^2 test, a significant value indicating that the two loci are linked.

The cross-over value The cross-over value between these two loci is $(c+d)/\text{total}$; this can be compared with the expected value of 34·6, again using χ^2.

	parental classes	cross-over classes	total
observed number (O)	a + b	c + d	a + b + c + d
expected number (E)	$\frac{65\cdot4}{100}$ × total	$\frac{34\cdot6}{100}$ × total	a + b + c + d
$(O{-}E)^2/E$			
χ^2_1			
probability			

Analysis of χ^2 In the table below are the results of five experiments as in the one just described. By computing a series of χ^2 values we can analyse these results and derive conclusions about the various factors which may influence the outcome of an experiment of this kind.

total	parental classes		cross-over classes		χ^2	degrees of freedom	proba-bility
	+	w m	w	m			
130	42	40	25	23	9·015	3	< 0·05*
161	51	40	34	36	4·292	3	> 0·05
200	56	41	49	54	2·680	3	> 0·05
57	23	7	12	15	9·456	3	< 0·05*
187	54	49	40	44	2·369	3	> 0·05
735	226	177	160	172	13·784	3	< 0·05*

*indicates a significant value of χ^2; that is, the data depart significantly from a 1:1:1:1 ratio.

If the two loci were not linked, if there were no differential survival of the four classes of offspring, and if all the five experiments gave similar results, then the total χ^2 would be very small. In this experiment the total $\chi^2 = 9\cdot015 + 4\cdot292 + 2\cdot680 + 9\cdot456 + 2\cdot369 = 27\cdot812$. Each of the five χ^2 values added to obtain this total is the result of testing the data for each experiment against an expected $1:1:1:1$ ratio. Since the total is quite large, any one or all three of the factors mentioned above may be influencing the results.

Linkage   If the loci were not linked, then there should be approximately equal numbers of parental and cross-over classes. In other words, the expected numbers in these two classes would be $735/2 = 367\cdot5$. However, the observed numbers are as follows:

$$\text{parental classes} = 226 + 177 = 403$$
$$\text{cross-over classes} = 160 + 172 = 332$$

Comparing the observed numbers with those expected we obtain a χ^2 value of $6\cdot858$ with 1 degree of freedom. This is significant at the 1% level and, therefore, the loci are linked. In other words, out of the total χ^2 of $27\cdot812$, $6\cdot858$ is caused by linkage.

Differential survival (viability)   In the bottom line of the table are the combined data for all five experiments. When these are tested against a $1:1:1:1$ ratio, a value of $\chi^2 = 13\cdot784$ (with 3 degrees of freedom) is obtained. Since we are here considering not only differences between parental classes and cross-over classes but also differences within these classes, the χ^2 value is partly caused by linkage and partly by viability differences. Since we already know the linkage χ^2 value, the viability χ^2 value will be $13\cdot784 - 6\cdot858 = 6\cdot926$. The degrees of freedom are obtained also by subtracting the linkage value from the total for the combined data. Thus, $3 - 1 = 2$. A χ^2 value of $6\cdot926$ with 2 degrees of freedom is significant at the 5% level, indicating that there is differential survival among the four classes. Obviously, if we wished to study only the linkage between these two loci it would be desirable to reduce this component as much as possible. This could be achieved by reducing crowding within the culture bottles, thus making it easier for the inferior classes to survive.

Heterogeneity of replicates   As stated in the previous section, the χ^2 that results from testing the bottom line of the table against a $1:1:1:1$ ratio is partly caused by linkage and partly by viability differences. It will not, however, have a component resulting from any differences (heterogeneity) between experiments because any such differences will be concealed when the individual experimental results are combined to form this bottom line of the table. The total χ^2 does contain a heterogeneity, linkage, and viability component and so the heterogeneity χ^2 value can be obtained by taking $13\cdot784$ from $27\cdot812$. The result is a heterogeneity χ^2 value of

14·028 with 12 degrees of freedom. This does not reach significance at the 5% level and, therefore, we can conclude that our replication is good. In other words, each of the five experiments gave results that were sufficiently alike to indicate that the outcome is reliable. Obviously, it is valuable to know that an experimental design is sound. Too much variation from one experiment to another will make it more difficult to draw conclusions from the data.

The complete analysis of χ^2 can now be set out:

source	χ^2	degrees of freedom	probability	significance
linkage	6·858	1	< 0.01	at 1% level
viability	6·926	2	< 0.05	at 5% level
total 1:1:1:1	13·784	3	< 0.05	at 5% level
heterogeneity	14·028	12	> 0.05	not significant
TOTAL	27·812	15	< 0.05	at 5% level

7 Natural selection against a mutant form

These experiments are best carried out either in milk bottles or in population cages. For the initial experiment the sex-linked recessive mutant white-eyed is used. Place a number of normal males and white-eyed virgin females in the experimental container. In a milk bottle about 20 flies (mixed sexes) will be enough; for a cage about 40 flies would be better. The population of flies can be examined once every generation (about 14 days). With the milk bottle all the adults can be etherized, removed, scored and counted before being returned to a bottle with fresh food. Similarly, the cage population can be removed, scored, counted, and returned. Fresh food tubes can be placed in the cage at regular intervals.

An alternative method for use with the cage is to sample the population at intervals and not to count every fly. Since cage populations have been known to build up to a thousand, this sampling procedure has a lot to recommend it. The danger is that during the sampling process errors might be introduced into the collected data which produce a false impression of the real situation. If you use a sampling technique, think carefully about possible sources of error.

A pooter can be used to remove a sample of about 200 adults. These are then examined and returned to the cage. If the cages or bottles are maintained for several months, the process of elimination of the mutant can be followed. Other mutants, such as vestigial wing and Bar eye, can be used and the rate of elimination compared. In addition, different cages can be subjected to various environmental conditions to see if any changes in the rate of elimination occur.

8 Competition between species

The experimental set up is identical with experiment 7 except that we now introduce into the containers two different species, *Drosophila melanogaster* and *D. simulans*, rather than two forms of one species.

The cages can be kept in a variety of temperatures: 15° C, 20° C, and 25° C, to study the effect of temperature on competition.

Repeat the experiment with *Drosophila melanogaster* and *D. funebris*. Different cages should contain food of different ages. In some, the food tubes should not be left in the cages for more than 2 or 3 weeks. In others, they can be left for a month or more. Finally, some cages should have both old and new food.

Other species pairs that might be used are *D. subobscura* and *D. obscura; D. phalerata* and *D. transversa;* and *D. melanogaster* and *D. busckii.*

9 Courtship behaviour

We can readily observe courtship behaviour in the laboratory by placing a pair of flies in a suitable observation chamber. The base of a small petri dish (50 mm × 10 mm) can be used containing *Drosophila* food medium, and covered with Cellophane held in position with a rubber band and pierced by a number of small pinholes for ventilation. To avoid condensation which would obscure the inside of the dish, place the Cellophane in position only when the apparatus is actually being used. After preparation the base can be stored until needed with the top half of the petri dish in position.

Note accurately the sequence and duration of each of the courtship movements and compare the results for different species. In *D. melanogaster*, observe the difference in courtship pattern when using a normal male and when using a male with the wings removed or when using a vestigial mutant.

10 Food preference

Larvae Use a large petri dish containing an agar surface as the experimental chamber. Offer four choices of food in the form of solutions deposited on the surface of the agar. Place these four drops equal distances apart in imaginary sectors of the dish and take care that they do not contaminate one another. Place about 20 feeding larvae in the centre of the agar, replace the petri-dish top and leave them for about 1 hour in a dark place. At the end of this period divide the agar into four sectors, one containing each food, and record the number of larvae in each. If the larvae show no preference for any one food then approximately a quarter should be in each sector. Compare the "expected" value of 5 with the observed number using the χ^2 test.

The foods used can include different varieties of yeasts (a great variety can now be obtained from shops that specialize in home wine-making), chemicals of the kind mentioned in Chapter 6, and fruit juices of various kinds. Do flies reared as larvae on a particular variety of food — apple, for example, show a preference for apple juice compared with other fruit juices?

Adults   A modified form of the sandwich box population cage described in Chapter 8 can be used to examine the food preference of adult flies. Apart from the ventilation hole in the lid only four other holes are made in the sandwich box, one in each side. Into these can be fitted, not the normal food tube, but a long test-tube. Place a different food in each and fit the tubes to the cage. Put 50 adult flies in the centre box and keep the whole apparatus in a dark place for 30 to 60 minutes. Then, quickly remove the tubes, stopper them with cotton-wool bungs, and count the number of flies in each exactly. The χ^2 test can be used as in the larval experiment.

Compare various species and their food preferences. Could these be related to the kind of habitat in which the flies occur naturally?

1   Field work

It is difficult to suggest anything more than a brief outline of the sort of work that can be attempted in the field because the details will depend on the time available, the number of students taking part, and the locality selected for study.

General survey   Select a number of habitats and make several collections in each. It is a good idea to begin field work with a survey because you will soon become familiar with a wide variety of species. If trapping sites within one habitat consistently yield a different mixture of species, a detailed study of the flora of the area (including fungi) and of various geographical features may suggest possible explanations.

One habitat throughout the year   The ideal situation would be to collect samples every week, but samples collected once a month will show interesting variations throughout the year. It is essential, as with all field investigations, to keep detailed notes on climatic changes and floral changes that may help to explain the variation in numbers of flies caught. A domestic garden is an ideal site for such a study.

Daily activity   Use cup traps and examine them every 1 or 2 hours, starting just before dawn and finishing just after sunset. As mentioned in

Chapter 3 activity patterns vary in different habitats so that similar studies can be carried out in different sites.

Vertical distribution in a woodland   Either cups or bottles can be used to trap the flies. If you use cups you will have to climb the tree to take the samples! If you use bottles, they can be suspended at the correct height by means of a piece of string tied round the neck of the bottle, passed over a branch, and pegged into the ground. The sample can then be retrieved by unpegging the string and lowering the bottle. However, because *Drosophila* do not enter bottle traps quite so readily as they do cups, it is wise to leave bottles in position for at least a week. The cups can be emptied several times a day.

Collection of fungi and other naturally occurring breeding sites   Collect and remove to the laboratory potential breeding sites and keep them individually in jars covered with fine nylon. Keep detailed notes, recording such things as species of fungus, date of collection and locality, temperature of the laboratory, and the dates of emergence of any *Drosophila* species.

Appendix

The mutants of *D. melanogaster*

A complete list of all the mutants of *D. melanogaster* is given by Lindsley and Grell (1967). The following list taken from this publication contains just a few of these mutants that are most useful. The figures enclosed in brackets indicate the chromosome on which the mutant is located and its position in terms of map units. For example, Bar is situated on chromosome 1 (the X-chromosome) and positioned 57 map units along its length. Note that several mutant characters can be present in one stock—wmB used in the three-point test-cross in the investigations, for example.

B   Bar (1–57·0). Eye restricted to a narrow vertical bar in the male and in the homozygous female. Heterozygous (+/B) females have a kidney-shaped eye.

b   black (2–48·5). Black pigment on the body and tarsi and along veins, darkening with age. Puparium usually somewhat lighter than in the normal type, and the newly emerged flies not clearly distinguishable from the normal type.

bw   brown (2–104·5). Eye colour light-brownish-wine on emergence, darkening to garnet. The red pigments of the eye are lacking. Adult testes and vasa colourless. Larval Malpighian tubules pale yellow. Produces white eyes in combination with v, cn, or st. (see Pm).

car carnation (1–62·5). Eye colour dark ruby. Body shape and proportions seem rounded. With st, eye colour is yellow-brown; with bw, brownish yellow to brown.

cn   cinnabar (2–57·5). Eye colour bright red, like v or st. Ocelli colourless. Eye colour darkens with age, but ocelli remain colourless. Larval Malpighian tubes pale yellow.

ct   cut (1–20·0). Wings cut to points and edges scalloped. Eyes smaller and somewhat kidney shaped. Abdominal bands warped. Antennae often deformed.

Cy   Curly (2–6·1). Wings curled upwards. Curvature caused by the unequal contraction of the upper and lower surfaces of the wing during the drying period following emergence from the pupal case. Usually homozygous lethal, but may emerge as dwarf with more extreme wing character.

D   Dichaete (3–40·7). Wings extended uniformly at 45° from body axis and elevated 30° above. Alulae missing. Head often deformed. Halteres turned down. Homozygous lethal.

dp   dumpy (2–13·0). Wings truncated and reduced to about two-thirds the length of those of the normal type.

e   ebony (3–70·7). Body colour shining black. Puparia much lighter than normal type. Classifiable throughout larval period by darkened colour of spiracle sheaths. Viability lowered to about 80% normal.

ey   eyeless (4–2·0). Eye reduced to three-fourths to one-half normal area, but varies from no eyes to extensive overlapping of normal type. Less extreme at low temperatures.

f   forked (1–56·7). Bristles shortened, gnarled and bent, with ends split or sharply bent. Hairs similarly affected, but this is visible only at high magnifications.

H   Hairless (3–69·5). Bristles, especially postverticals and abdominals, missing. Veins IV and V do not reach wing margin; occasionally true of II also. Eyes larger than normal type; body colour somewhat paler. Homozygous lethal.

l( ) lethal (). General term used to describe recessive mutations that lead to death of most or all homozygous carriers. The symbol l is followed by the chromosome upon which the lethal is situated and then by the name or number of the particular mutant. For example, l(3)blo-1 means lethal (3) bloated larvae. Larvae become very large and transparent and die in the pre-pupal stage.

L   Lobe (2–72·0). Heterozygous L eyes slightly smaller, with nick in anterior edge, and lower half of eye reduced more than upper;

132

overlaps normal type. Homozygous L eyes much smaller and less variable. Several allelomorphs of Lobe are known. For example, L^2, which has a more extreme effect. Eyes of $L^2/+$ as small as or smaller than L/L above. L^2 homozygotes have tiny eyes and are poorly viable or completely lethal.

ı miniature (1–36·1). Wing size reduced; only slightly longer than abdomen and with normal proportions. Wings dark grey and less transparent than normal.

m Plum (2–104·5). Name sometimes given to one of the allelomorphs of bw, bw^{vi} (brown-Variegated). Eye colour like bw, mottled with darker spots that deepen in red colour with age. With st or v, has pale orange ground with dark orange spots. Homozygous Pm is lethal; heterozygotes fully viable and fertile.

sable (1–43·0). Body colour dark. Classification good at 19°C; overlaps normal type increasingly with higher temperatures. Viability sometimes reduced.

b Stubble (3–58·2). Bristles of Sb/+ less than one-half normal length, and somewhat thicker than normal type. Homozygous lethal.

cx Extra sex comb (3–47·0). Sex combs may be present on all six legs of male. At least one extra sex comb present in 75–90% of males. Third pair of legs less often affected than second. Homozygous lethal.

e sepia (3–26·0). Eye colour brown at emergence, darkening to sepia and becoming black with age. Pigmentation of ocelli normal.

s spineless (3–58·5). Bristles only a little larger than hairs; dorso-centrals least reduced. Postscutellars erect. No effect on legs or aristae.

t scarlet (3–44·0). Eyes bright vermilion, darkening with age. Ocelli colourless, even in old flies. Eyes of the double recessive st/st bw/bw are white.

vermilion (1–33·0). Eye colour bright scarlet owing to absence of brown pigment. Ocelli colourless. The combination v/v bw/bw has white eyes.

g vestigial (2–67·0). Wings reduced to vestiges; usually held at right angles to body. Halteres also reduced. Viability somewhat reduced.

w   white (1–1·5). Eyes pure white. Ocelli, adult testes sheaths, and larval Malpighian tubules colourless. There are many allelomorphs of white; two are given below.
w^a white-apricot. Eye colour of male yellowish-pink; female eye colour somewhat yellower.
w^e white-eosin. Eyes of female yellowish-pink, male and w^e/w^e female lighter.

y   yellow (1–0·0). Body colour yellow; hairs and bristles brown with yellow tips. Wing veins and hairs yellow. Larval mouthparts yellow to brown, hence distinguishable from the dark brown of normal type.

BIBLIOGRAPHY

Baker, H. G.   1965   The genetics of colonising species. Academic
Stebbins, G. L. (Eds.)   Press, New York.

Basden, E. B.   1953   The vertical distribution of Drosophilidae in
Scottish woodlands. *Drosophila Information Service** (see page 139) **27**, 84.

Basden, E. B.   1954*a*   The distribution and biology of Drosophilidae
(Diptera) in Scotland, including a new species
of *Drosophila. Transactions of the Royal Society of Edinburgh* **62**, 602–654.

Basden, E. B.   1954*b*   Diapause in *Drosophila* (Diptera: Drosophilidae). *Proceedings of the Royal Entomological Society of London (A)* **29**, 114–118.

Beardmore, J. A.   1967   *Drosophila andalusiaca*, a polymorphic species
new to Holland. *Archives Neeland aises de Zoologie* **17**, 275–277.

Bennet-Clark,   1968   The wing mechanism involved in the court-
H.C. &   ship of *Drosophila. Journal of Experimental*
Ewing, E. W.   *Biology* **49**, 117–128.

Bennet-Clark,   1970   The love song of the fruit fly. *Scientific*
H. C. & Ewing, E. W.   *American* **223**, 84–92.

Burla, H.   1951   Systematik, Verbreitung und Oekologie der
Drosophila Arten der Schweiz. *Revue suisse de Zoologie* **58**, 23–175.

Carson, H. L. &   1948   Reproductive diapause in *Drosophila robusta.*
Stalker, H. D.   *Proceedings of the National Academy of Science* **34**, 124–129.

Carson, H. L. &   1951   Natural breeding sites for some wild species
Stalker, H. D.   of *Drosophila* in the eastern United States.
Ecology **32**, 317–330.

Castle, W. E., 1906 Carpenter, F. W., Clark, A. H., Mast, S. O., & Barrows, W. M.
The effects of inbreeding, cross-breeding and selection upon the fertility and variability of *Drosophila*. *Proceedings of the American Academy of Arts and Science* **41**, 729–786.

Demerec, M. & 1967 Kaufmann, B. P.
Drosophila guide. Carnegie Institution of Washington, Washington, D.C.

Dobzhansky, Th. 1951
Genetics and the origin of species. Columbia University Press, New York.

Dobzhansky, Th. 1970
Genetics of the evolutionary process. Columbia University Press, New York.

Dobzhansky, Th., 1956 Cooper, D. M., Phaff, H. J., Knapp, E. P., & Carson, H. L.
Studies on the ecology of *Drosophila* in the Yosemite region of California: IV Differential attraction of species of *Drosophila* to different species of yeasts. *Ecology* **37**, 544–550.

Dobzhansky, Th. 1944 & Epling, C.
Contributions to the genetics, taxonomy and ecology of *Drosophila pseudoobscura* and its relatives. *Carnegie Institution of Washington Publication* **554**, 1–46.

Dobzhansky, Th. 1950 & Pavan, C.
Local and seasonal variations in relative frequencies of species of *Drosophila* in Brazil. *Journal of Animal Ecology* **19**, 1–14.

Dyson-Hudson, 1954 V. R. D.
The taxonomy and ecology of the British species of *Drosophila*. D.Phil. thesis (deposited in the Bodleian Library, Oxford, and in the library of the Department of Zoology and Comparative Anatomy, Oxford University).

Dyson-Hudson, 1956 V. R. D.
The daily activity rhythm of *Drosophila subobscura* and *D. obscura*. *Ecology* **37**, 562–567.

Ewing, A. W. & 1968 Bennet-Clark, H. C.
The courtship songs of *Drosophila*. *Behaviour* **31**, 288–301.

Ferris, G. F. 1950
In "Biology of *Drosophila*" edited by M. Demerec; John Wiley and Son Inc. New edition (1965) by the Hafner Publishing Company, New York.

Fonseca, E. C. M. 1965 A short key to the British Drosophilidae
d'Assis   (Diptera) including a new species of *Amiota*.
              *Transactions of the Society for British Ento-
              mology* **16**, 233–244.

Frydenberg, O. 1956 The Danish species of *Drosophila* (Dipt.).
              Entomologiske Meddelelser **27**, 249–294.

Gordon, C.   1942 Natural breeding sites of *Drosophila obscura*.
              Nature, London **149**, 499.

Hadorn, E.,   1952 Beitrag zur Kenntnis der Drosophila-Fauna
Burla, H.,   von Südwest-Europa. *Zeitschrift für indukt
Gloor, H., &   Abstammungs und Vererbungslehre* **84**, 133–
Ernst, F.   163.

Heed, W. B., & 1965 Unique sterol in the ecology and nutrition of
Kircher, H. W.   *Drosophila pachea*. *Science* **149**, 758–761.

Howard, L. O. 1900 A contribution to the study of the insect
              fauna of human excrement (with special
              reference to the spread of typhoid fever by
              flies). *Proceedings of the Washington Academy
              of Science* **2**, 541–604.

Kaufmann, B. P. 1939 Induced chromosome rearrangements in
              Drosophila melanogaster. *Journal of Heredity*
              30, 179–190.

Laurence, B. R. 1953 Some Diptera bred from cow dung. *Ento-
              mologists' Monthly Magazine* **89**, 281–283.

Lewontin, R. C. 1959 On the anomalous response of *D. pseudo-
              obscura* to light. *American Naturalist* **93**,
              321–328.

Lindsley, D. L. 1967 Genetic variations of *Drosophila melano-
& Grell, E. H.   gaster*. *Carnegie Institution of Washington
              Publication* 627.

Miller, A.   1950 In "Biology of *Drosophila*" edited by M.
              Demerec; John Wiley and Son Inc. New
              edition (1965) by the Hafner Publishing
              Company, New York.

Monclus, M.   1964 Distribucion u ecologia de drosofilidos en
              Espana. *Genetica Iberica* **16**, 143–165.

Okada, T.   1962 Bleeding sap preference of the drosophilid
              flies. *Japanese Journal of Applied Entomology
              and Zoology* **6**, 216–229.

Patterson, J. T.   1943   The Drosophilidae of the southwest. *University of Texas Publication* **4313**, 7–216.

Patterson, J. T.   1952   Evolution in the genus *Drosophila*. Mac-
& Stone, W. S.   millan, New York.

Pavan, C.,   1950   Diurnal behaviour of some neotropical species
Dobzhansky, Th.,   of *Drosophila*. *Ecology* **31**, 36–43.
& Burla, H.

Prevosti, A.   1966   Chromosomal polymorphism in western
                              Mediterranean populations of *Drosophila
                              subobscura*. *Genetical Research* **7**, 149–158.

Rasmuson, B. &   1969   *Drosophila* species in the northern part of
Johansson, H.   Sweden. *Drosophila Information Service** (see
                              page 139) **44**, 188.

Reed, S. C. &   1949   Natural selection in laboratory populations
Reed, E. W.   of *Drosophila*: II Competition between a
                              white eye gene and its wild type allele.
                              Evolution **4**, 34–42.

Shaw, M. W.   1968   *Drosophila* species (Dipt. Drosophilidae) as-
                              sociated with pea silage. *Entomologists'
                              Monthly Magazine* **104**, 236.

Shorrocks, B.   1969   A note on *Drosophila* species along the Tyne
                              valley. *The Entomologist* **102**, 229–230.

Shorrocks, B.   1970   A note on *Drosophila* species at Woodchester
                              Park, Gloucestershire. *The Entomologist* **103**,
                              286–288.

Sinnot, E. W.,   1958   Principles of genetics. McGraw-Hill Book
Dunn, L. C., &   Co., New York.
Dobzhansky, Th.

Sobels, F. H.,   1954   The distribution of the genus *Drosophila* in
Vlijm, L., &   the Netherlands. *Archives Neerlandaises de
Lever, J.   Zoologie* **10**, 357–374.

Spencer, W. P.   1950   In "Biology of *Drosophila*" edited by M.
                              Demerec; John Wiley and Son Inc. New
                              edition (1965) by the Hafner Publishing
                              Company, New York.

Unwin, E. E.   1907   The vinegar fly (*Drosophila funebris*). *Trans-
                              actions of the Entomological Society of London
                              part 1*, 285–302.

Drosophila Information Service is an approximately annual publication (latest issue number 46, 1971) prepared at the Department of Biology, University of Oregon, Eugene, Oregon, U.S.A. It contains details of the mutants of *Drosophila melanogaster* and other species, and a list of wild species maintained by laboratories all over the world. A directory of these laboratories and their personnel is included. In addition, research notes and techniques in handling and keeping *Drosophila* are included. Anyone interested in further details should write to the Biology Department at Oregon.

INDEX

ACKNOWLEDGEMENTS

Thanks are due to the undermentioned authors and publishers for permission to redraw various illustrations from their publications:
to G. F. Ferris for Figures 5B and C, and 7C and D.
A. Miller for Figures 9, 10, 11, 12, and 13.
D. Bodenstein for Figures 15C, and
W. P. Spencer for Figure 51 which were redrawn from illustrations appearing in 'Biology of *Drosophila*', edited by M. Demerec, and published in a new edition in 1965 by the Hafner Publishing Company Inc., New York.
to H. C. Bennet-Clark and E. W. Ewing for Figures 6 and 31 which were redrawn from their paper in the *Journal of Experimental Biology*, **49** (1968) 117–128, published by the Cambridge University Press.
to Th. Dobzhansky and C. Pavan for Figure 19 which was redrawn from their paper in the *Journal of Animal Ecology*, **19** (1950) 1–14, and reproduced by kind permission of Blackwell Scientific Publications, Oxford.
to V. R. D. Dyson-Hudson for Figures 21, 40D, 41C, 42C, and 43E and F, which were redrawn from her paper in *Ecology*, **37** (1956) 562–567, and reproduced by kind permission of the Duke University Press, North Carolina, U.S.A.
to E. W. Sinnot, L. C. Dunn, and Th. Dobzhansky for Figure 23 which was redrawn from their book 'Principles of Genetics', published by the McGraw-Hill Book Company, New York.
to S. C. and E. W. Reed for Figure 24 which was redrawn from their paper in *Evolution*, **4** (1949), 32–42 and reproduced by kind permission of the Department of Entomology, University of Kansas, U.S.A.
to B. P. Kaufmann for Figure 25 which was redrawn from his and M. Demerec's '*Drosophila* guide', published by the Carnegie Institution of Washington, and reproduced by kind permission of the publishers.
to A. Prevosti for Figure 29 which was redrawn from his paper in *Genetical Research*, **7** (1966) 149–158, published by the Cambridge University Press.
to H. Burla for Figures 37B and C which were redrawn from his paper in *Revue suisse de Zoologie*, **58** (1951) 23–175.

The publishers are also most grateful to Dr H. C. Bennet-Clark of the Department of Zoology, University of Edinburgh, for the photographic print reproduced in Figure 30, and to Professor Berwind P. Kaufmann, of the Department of Zoology, University of Michigan, for the photomicrographs reproduced in Figures 26 and 28.

The table which appears on page 118 is taken from Table IV of Fisher and Yates's "Statistical Tables for Biological, Agricultural and Medical Research", published by Oliver and Boyd.